Georg Zurlo

Cichlids

Purchase, Care, Feeding, Diseases, Behavior, and Breeding

Drawings by Fritz W. Köhler

Consulting Editor: Matthew M. Vriends, PhD

The title of the German book is *Buntbarche Cichliden*

Translated from the German by Rita and Robert Kimber

All inquiries should be addressed to:
Barron's Educational Series, Inc.
250 Wireless Boulevard
Hauppauge, NY 11788

Library of Congress Catalog Card No. 90-19430

International Standard Book No. 0-8120-4597-1

Library of Congress Cataloging-in-Publication Data

Zurlo, Georg.
(Buntbarche Cichliden. English]
Cichlids : purchase, care, feeding, diseases, behavior, and breeding / Georg Zurlo : drawings by Fritz W. Köhler; Consulting editor: Matthew M. Vriends.
p. cm.
Translation of: Buntbarche Cichliden.
Includes bibliographical references.
Includes index.
ISBN 0-8120-4597-1
1. Cichlidae. 2. Aquarium fishes. I. Title.
SF458.C5Z8713 1991 90-19430
639.3'758—dc20 CIP

Printed in Hong Kong
5 6 7 8 4900 11 10 9 8 7

The color photos on the covers show:
Front cover: *Cyphotilapia frontosa* (Frontosa); 3 females.
Inside front cover: *Hemichromis cristatus* (Crown Jewel Cichlid or "Red Cichlid") with fry.
Inside back cover: *Steatocranus casuarius* (Lionhead, Blockhead, or Lumphead).
Back cover: *Biotodoma cupido* "Santarem" (Cupid Cichlid).

Photo credits
Kahl: front cover, pages 18 (above right, middle, below left and right), 45 (below right), 56, inside back cover.
Linke: inside front cover, pages 17, 27, 46.
Meulengracht-Madson/Biofoto: page 53.
Stawikovski: pages 18 (above left), 28 (above), 45 (above left and right), back cover.
Werner: page 45 (middle and below left).
Zurlo: page 28 (below).

About the author

Georg Zurlo is a teacher who has been keeping fish for many years. He is the author of many articles. Cichlids of the New and Old World are his field of special interest.

A note of warning

In this book, electrical equipment commonly used with aquariums is described. Please be sure to observe the safety rules on page 15; otherwise there is a danger of serious accidents.

Before buying a large tank, check how much weight the floor of your apartment can support in the location where you plan to set up your aquarium (see page 15).

Sometimes water damage occurs as a result of broken glass, overflowing, or a leak in the tank. An insurance policy that covers such eventualities is therefore highly recommended.

Make sure that children (and adults) do not eat any aquarium plants. These plants can make people quite sick. Also make sure that fish medications are out of reach of children (see page 33).

Contents

Contents

Preface

Many aquarists, upon seeing a tank with gorgeously colored cichlids, develop a great enthusiasm and hankering for these fish. But cichlids have a reputation that is not altogether encouraging: They are said to be bullies that not only attack other fish but also uproot and devour plants and keep plowing through the sand in the bottom of the tank. However, most of the bad experiences people have had with cichlids are the result of improper care and conditions. Either the cichlids were housed in tanks that were too small for them, or they were combined with other fishes that were not compatible.

This pet owner's guide will tell you how to avoid mistakes of this sort from the very outset. Georg Zurlo has been keeping and breeding different kinds of cichlids for many years, and in this book he offers advice and recommendations on how to treat these fish properly. His suggestions are easy to follow—even for beginners. His tips for the proper selection and combination of various species and his concrete information on filtering, water quality, and setting up a tank will prevent possible disappointment to any hobbyist about to try his or her luck with cichlids.

Appropriate nutrition plays an especially important role for cichlids. True, the majority of cichlids are omnivorous, but the food they get has to be of high quality and the diet varied if the fish are to thrive properly. That is why the author explains in detail how the different species should be fed.

Although cichlids rarely get sick if they are kept and cared for properly, diseases do sometimes crop up. The chapter, Diseases of Cichlids, offers information and advice for such cases.

Because of their fascinating behavior, cichlids have become a favorite subject of study for ethologists. For instance, cichlids can change color within seconds, which they do to signal various moods to others of their kind—moods like aggressiveness or the desire to stay at a safe distance, or, in a different context, their readiness to spawn. In the chapter devoted to this topic you will find more on the behavior of cichlids as well as some suggestions of what to watch for in your own cichlids. The reproductive behavior of cichlids is also extremely interesting. Almost all species engage in some definite form of brood care. The so-called mouthbrooders among them keep the eggs inside the mouth to hide them from predators. Even after the fry have hatched, they still are allowed to swim into their parents' mouths when danger lurks. In the chapter on breeding cichlids, Georg Zurlo shares information and advice based on his own experience and actual practice.

In the last section of the book you will find detailed descriptions of cichlids from all over the world. These descriptions include size, appearance, sexual differences, behavior, and specific directions on how to keep and breed the various species. The author has concentrated especially on fishes that can be recommended for beginners, and the information given here will help you choose and combine the right cichlid species.

The author sometimes puts some of the genus names in quotes (''Geophagus,'' ''Haplochromis''). These are old designations that still are used in the trade, though they no longer are accepted by taxonomists.

Lifelike color photos—taken by top photographers in the field of aquarium photography—together with informative drawings convey an impressive picture of the color variety and the downright enthralling behavior of cichlids. The author of the book and the editors of Barron's pet owner's manuals and handbooks wish you a lot of fun with your cichlids.

The author and the publisher wish to thank all those who had a share in producing this book: Michael Prädel, Ernst Sosna, and Uwe Werner, for their helpful support; Ulrich Schliewen, for his professional advice; the photographers, for their exceptional color photos; the illustrator, Fritz W. Köhler, for his informative drawings; and Harald Jes, director of the Aquarium at the Cologne Zoo, for checking the chapter, Diseases of Cichlids.

Important Facts About Cichlids

The Cichlid Family

Most cichlids are characterized by beautiful colors and markings. Even within the same species different shades and patterns can develop, a phenomenon that is called polymorphism.

There is such a variety of shapes and sizes among the many cichlid species that even experienced fanciers of aquarium fishes often are surprised. Thus, the "giants" within the family measure from 28 to 32 inches (70–80 cm), whereas the smallest members are fully grown at as little as 1.4 inches (3.5 cm).

The body shapes (see drawing on right) of various species also vary tremendously. Most cichlids have a typical slender fish shape, but in different species this shape can vary from "normal" to "pencil-shaped." A few species are high-backed, that is, they have steeply arched backs; and some, like the popular discus fishes, are almost completely round and flat.

The cichlid family takes in over a thousand different species. This book is geared primarily to the concerns of aquarists new to the art of keeping cichlids. That is why, in the descriptions of popular cichlids that starts on page 49, you will find listed mostly those species that are suitable for beginners. The many possible exceptions to the rules for the care of fish given in this book—exceptions that apply to certain harder-to-please species—are ignored here.

Why the Latin Names Are Important

Many cichlid species simply have no popular English names. That is why any cichlid fancier should make an effort to master the scientific names of his or her fish. If you know the Latin names, you can refer to your cichlids without the possibility of being misunderstood, and you'll have no problems communicating with aquarists in other countries. In cases where the English name of a species is used commonly and universally, this name also will be used in the descriptions of popular species starting on page 49.

Carl Linné introduced the so-called binomial nomenclature in 1766, and ever since then every organism—whether plant or animal—has been classified according to this system.

The *first word* of the name tells what genus the organism belongs to, and the *second word* indicates the species.

In addition, the *place of origin* sometimes is added in quotation marks because, in the case of some cichlids, strains from different geographic locations may vary in coloration.

Let us look at the name *Pelvicachromis taeniatus* "Kumba" as an example. This is the scientific name of the striped kribensis, a cichlid that belongs to the genus *Pelvicachromis* and the species *taeniatus* and that comes originally from the town of Kumba in western Cameroon.

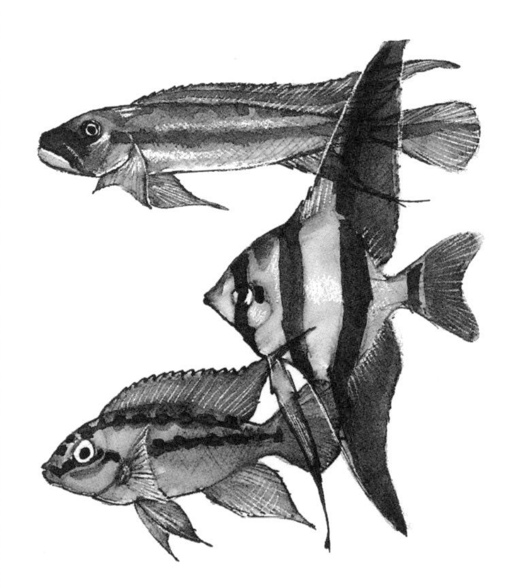

Different body shapes of cichlids. Cichlids can be long and pencil-shaped, laterally flattened with wide fins, or even round like a disk.

Important Facts About Cichlids

Where Cichlids Come From

This fish family is found only in the tropical and subtropical zones of three continents.

Africa: Especially in the lakes of the Great Rift Valley in East Africa, such as Lakes Tanganyika and Malawi, a great variety of cichlid species and some very interesting feeding specializations have developed. There one finds cichlids that "graze" on the algae growing on rocks (the so-called *aufwuchs* eaters, see page 25), and there are even some that bite off and eat the scales of other fish. Other African lakes as well as West African rivers, such as the Congo—now called Zaire River—also have cichlids of many different genera, species, and color varieties.

The Americas: The Amazon watershed with its many rivers is undoubtedly the most important source of cichlids in this part of the globe. In addition to innumerable other freshwater fishes, many exceptionally colorful and interesting cichlid species are found here.

Only a small number of species live in the subtropical areas of the American continents, as, for instance, in Texas.

Asia: Two species of the genus *Etroplus* have become established in India and on Ceylon.

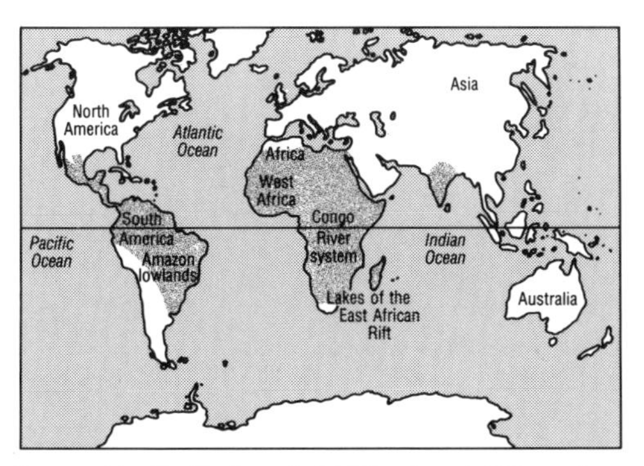

Geographic distribution of cichlids. Over 1,000 species of cichlids are found in Africa, Central and South America, and Asia.

The Natural Environment of Cichlids

Cichlids occupy very different habitats, and the nature of a particular habitat to a large extent determines the behavior, feeding habits, body shape, and coloration of the species that have adapted to it.

The following are typical cichlid biotopes:

- loose rock and rocky shorelines;
- sandy bottom and muddy bottom; and
- shores with and without vegetation.

Rocky shorelines and loose-rock bottom: Numerous cichlids live in areas with loose rock. Here the bottom is covered with gravel and stones (ranging from fist size to small blocks). Lake Tanganyika especially has stretches of shoreline of this type, and the edges of many rivers in Central America are like this. The loose-rock shorelines of Lake Tanganyika offer good cover to cichlids— such as *Eretmodus cyanostictus*—living in those waters, and aufwuchs-eating species (see page 25) "graze" on the algae that grow on the rocks. It is not difficult to recreate this type of environment in an aquarium. Use a tank with a large bottom area. Cover the bottom with sand and distribute lots of stones of different sizes on it (aquarium types 6 and 8, see page 13).

A rocky shoreline is the underwater zone of steep, rocky banks, such as are found primarily on lakes Tanganyika and Malawi. Such a shoreline is characterized by big rocks piled on top of each other. The way the rocks have come to rest creates innumerable caves of various sizes, into which the cichlids can retreat when they sense danger. This type of littoral zone is the typical habitat of algae-eating mouth brooders (see page 25), such as the cichlids of the genus *Tropheus*. You can create a similar environment in an aquarium by piling rocks on top of each other (aquarium types 6 and 8, see page 13).

Sandy bottom and muddy bottom: Both Lake Tanganyika and Lake Malawi have large sandy parts. The sand extends over wide areas with only occasional rocks or plants here and there. Often, empty snail shells dot the bottom and in some places collect in what could be called snail cemeteries. In Lake Tanganyika, cichlids of the genus *Neolamprologus* inhabit these snail shells. This environment, too, can be recreated quite easily in an aquarium (aquarium type 9, see page 13).

Muddy bottoms usually occur at inlets where rivers enter a lake and where floating particles carried by the moving water settle. Here, too, cichlids, such as *Triglachromis otostigma*, are found.

Because no aquarist is likely to want muddy water in his or her tank, cichlids from this kind of habitat usually are kept in tanks with a gravel or sand bottom (aquarium types 3 and 8, see pages 12–13).

Shores with and without vegetation: Shores with dense vegetation and small, shallow bodies of water filled with plant growth also are environments that cichlids have adapted to. Several *Pelvicachromis* species, for instance, use such regions as a regular habitat and as an area of retreat.

Some suggestions for planting an aquarium for cichlids are found on page 20 and under aquarium types 1 and 2 on page 12. In rivers and lakes without littoral vegetation, tree roots hanging into the water and fallen trees or tree limbs offer protection to cichlids. This kind of littoral zone is found along many stretches of the large South American rivers. Discus and angelfish come from this kind of environment. You can imitate this kind of shore without plant life by placing in your aquarium bizarrely-shaped pieces of bog wood (see aquarium types 3 and 4, page 12). Any wood used in fish tanks has to be dead and preferably aged in the acidic soils and waters of bogs, so that there is nothing left to rot.

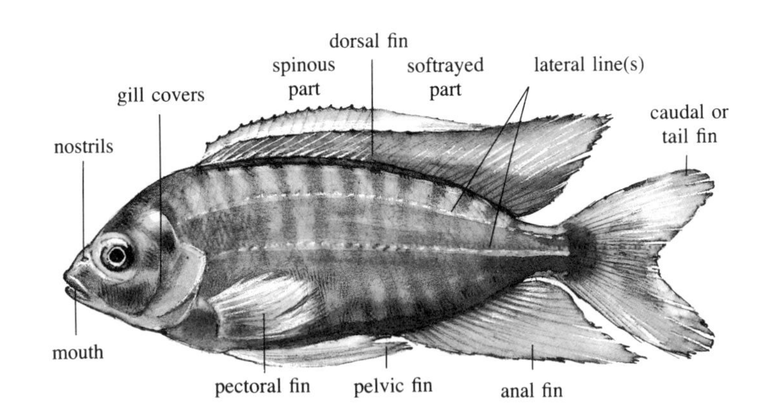

The body of a cichlid. Familiarity with the anatomy of cichlids can help you identify species you don't yet know.

Advice for Buying

What to Watch Out For

Hobbyists without experience in keeping cichlids should look for species with the following qualities: The fish grow to no more than 3 or 4 inches (8–10 cm) and are nonaggressive (toward their own kind as well as toward other species in the tank). They don't need too much space in an aquarium, and their requirements for water quality are moderate. Finally, they do not attack and eat other fish.

Note: You will find detailed information on species suitable for beginners on pages 49–67.

Where You Can Buy Cichlids

Pet dealers usually have a good selection of the most popular kinds of cichlids.

Private fish breeders often raise some rarer kinds of cichlids. You can find names of breeders in the classified section of aquarists' magazines (see Addresses and Bibliography, page 68). Or you can go to a meeting of a local aquarists' club or to a show organized by the American Cichlid Association. You may be able to purchase fish there, or the organizers of the event will be able to give you addresses of breeders in your area.

Which Fish to Buy

Buy fish that are still immature. Even though the full color has not yet developed in the juveniles of most cichlid species, buying young fish offers a number of advantages. Immature fish are less expensive than mature ones, and they adjust more easily to a new aquarium. You yourself are involved in raising the fish and thus can see to it that your cichlids will develop into healthy and hardy fish. You will have to wait longer to breed the fish (see Breeding, page 40) if you begin with juveniles, but you are less likely to run into problems, and chances for success are greater than if you try to breed fish bought full grown.

Is This Cichlid Healthy?

Many cichlids offered for sale come from excellent local fish breeders. But the "most popular" species often are mass-produced in huge hatcheries in Asia, which may make the price of the fish attractive but has a less positive effect on their health. When buying aquarium fish, you never should be swayed by a low price but always base your choice on the state of health of the fish.

You will be able to judge for yourself how healthy the fish are if you keep in mind the following as you watch them:

- Cichlids should be eager eaters. Ask to watch a demonstration feeding.
- Healthy fish display no shyness even in a dealer's tank. Only wild catches are an exception, that is, fish that were caught in their native waters and were not bred in a tank.
- The fish should not swim around with their fins clamped to the body, and they should not have misshapen heads, spines, or fins.
- Don't buy fish with damaged fins or skin — the result of being bitten by other fish in the tank.

S Fish with spots or a film on the skin, which are caused by diseases like white spot (see page 34) or by fungi (see page 34), are sick.

- Cichlids whose excreta trail after them in long, slimy white threads are suffering from serious intestinal disease.

My tip: Watch the behavior of cichlids in a dealer's tank over an extended period of time. This is the only way to get a reliable impression of the fish's state of health.

Telling the Sexes Apart

For an inexperienced aquarist, telling the males from the females in a shoal of young cichlids may be difficult if not impossible. However, sexing fully grown fish often is quite easy because the sexes may differ in size, coloration, body shape, and in the appearance of the genital papil-

Mouthbrooders. Mouthbrooding cichlids pick up the eggs in their mouths—usually the female does this—and keep them there until they hatch.

la. More detailed information on this subject is given in the descriptions of individual cichlid species starting on page 49.

With immature cichlids, size often may give a clue as to the sex of the fish, assuming, of course, that the entire shoal is from the same spawning. In such a case, the larger individuals are usually males and the smaller ones, females.

Should You Get Pairs or Small Groups?

Most cichlids are decidedly social. That is why you always should buy at least a pair of any one species. Keeping small groups, that is, five or more fish of the same species, is even better. Having just one cichlid in a tank is out of the question. A cichlid is kept singly only in an emergency situation, as when you have to isolate an aggressive individual.

Pairs: You sometimes can find pairs of a cichlid species that appeal to you with the pair bond already formed in the dealer's tank. You can tell that this is the case if, when watching them for a while, you see them defend a certain spot against intruders. Sometimes the pair already will have started a nest in this spot.

If you are unable to spot such a pair, try to find at least one male and one female of the same species. A knowledgeable pet dealer or a breeder can help you select the fish.

Groups: Buy at least five young fish of the same species. You should have more females than males because, in many species, males are considerably more aggressive than females. Individual females are attacked less and suffer less permanent stress if the males' aggressiveness is spread over a larger number. Three females to two males is a good ratio to start out with.

Combining Different Cichlid Species

If you would like to keep several kinds of cichlids together, you have to compare the characteristics and requirements of the different species as indicated in the descriptions starting on page 49. Make very sure that the species you want to combine

- do not have the same territorial needs,
- differ in their reproductive behavior,
- do not look alike,
- are not exceptionally aggressive,
- are of approximately the same size,
- have compatible feeding habits, and
- require the same water quality.

Note: Wrong combinations can give rise to serious fighting and even to outright killing. In addition, poorly matched fish are under constant stress, which is bound to affect their health.

Territorial needs: In nature, all cichlids are oriented toward the water's bottom; they are bottom dwellers, though not all of them live equally close to it. That is why you should combine fish that prefer the bottom of the tank with ones that like higher areas among plants or rock piles. You can assign different species to different levels in the aquarium. An example of a good combination of cichlids from West Africa is a community made up of *Steatocranus casuaris* and *Anomalochromis thomasi*. The former inhabit exclusively cavities very close to the bottom, whereas the latter usually stay "one floor up," in the middle range of the tank. Another possibility of preventing territorial competition is to create more than one habitat in a tank. Planted areas, rocky areas, and sandy areas are each preferred by different cichlids. Or, finally, you can organize separate territories when you set up the tank by piling up stone walls, building wooden barriers, or creating other structures.

Brooding behavior: Another successful way of achieving a compatible community is to combine fish that spawn in open water with ones that spawn in cavities—the so-called secret brooders—and with mouthbrooders (see Breeding Cichlids, page 40). If you want to have African dwarf cichlids, for instance, you can combine the open-water spawning *Anomalochromis thomasi* with the mouthbrooder *Pseudocrenilabrus multicolor* and add a secret brooder from the genus *Pelvicachromis*. This way you will at least be able to prevent competition over spawning sites.

Appearance: Make sure you don't combine species that have similar coloration and markings (see page 37). If you go against this rule, serious fights may erupt, especially between rival males.

Size: Differences in size should not be so big that the larger fish eat the smaller ones, chase them to exhaustion, or in any other way contribute to permanent stress.

Aggression: Neophyte aquarists should not combine unusually belligerent species. Information on the aggressiveness of different cichlids is given in the descriptions starting on page 49.

Feeding habits: Most cichlids are omnivorous (see page 25). But there are some that have specialized in certain foods and that should not be kept together with fish whose food habits are different. For example: Typical plant or algae eaters, such as the *Tropheus* species, develop potentially fatal intestinal problems if they are fed a low-fiber meat diet. On the other hand, a diet based on bulky plant matter leaves carnivorous fish perpetually hungry.

Water: Combine only cichlids that have identical requirements for water quality.

Note: Some suggestions for successful cichlid communities are given in the table on pages 12 and 13.

Combining Cichlids with Other Kinds of Fish

If you would like to combine your cichlids with members of other fish families, you have to select "companion fish" that are assertive and can put up with adverse conditions. To be able to get out of a cichlid's way quickly if necessary, they should be agile swimmers, and they must not be slow and tentative eaters, for otherwise they will go hungry or even starve to death in a tank with cichlids. The decision of whether the cichlids or the other fish should be the major focus of the aquarium is up to you. You can place a single pair of cichlids in a community tank with other fishes, or you may want to have a tank full of cichlids with just one or two catfish added.

Some fishes that have proven in actual practice to be compatible with cichlids are listed below.

African and South American Tetras

Genera/species: All the larger tetras, such as Congo tetras (*Phenacogrammus*) and *Hyphessobrycon* or *Hemigrammus* species measuring over 1½ inches (4 cm), like the bleeding heart tetra (*Hemigrammus erythrostigma*). Also cardinal tetras (*Cheirodon axelrodi*), neon tetras (*Paracheirodon innesi*), and the false rummynose (*Petitella georgiae*).

Life pattern: Shoal fish (keep at least six) for tanks at least 32 inches (80 cm) long.

Feeding: Live, frozen, and dry food.

Advice for Buying

Model Communities

Aquarium type/size	Interior	Water	Occupants
1. Small, planted community tank for cichlids from South America or West Africa. 24 to 32 inches (60–80 cm)	Dense planting around edges, perhaps some low plants in foreground; small rock structures; wood decorations; bottom material a mixture of sand and gravel (50:50) with diameter of up to ⅜ inch (8 mm).	ph 7 or less, hardness up to 12° dH, peat filtering recommended.	No more than two pairs of West African and/or South American dwarf cichlids: *Anomalochromis thomasi* and *Pelvicachromis pulcher*, or a *Steatocranus* species or *Papiliochromis ramirezi* and *Nannacara anomala*, or *Laetacara curvicps* and an *Apistogramma* species.
2. Large, planted community tank for cichlids from South America or Africa. 36 to 48 inches (90–120 cm), possibly up to 60 inches (150 cm); if larger, very time consuming (plant care!)	Same as under 1. but the ratio between sand and gravel doesn't matter.	Same as under 1.	Up to four pairs of West African and/or South American dwarf cichlids (see aquarium type 1) plus one pair of medium size, nonaggressive West African or South American cichlids, such as *Chromidotilapia guntheri*, *Teleogramma brichardi* and *Tilapia joka*, or *Aequidens maronii* and *Papiliochromis altispinosa*, or "*Geophagus*" *steindachneri* and *Pterophyllum scalare*, or four to five discus fish and four to five *P. scalare* or *P. altum*.
3. Small community tank without plants for Central American cichlids. At least 40 inches (100 cm)	Flat rocks leaned against the back wall, rock piles, decorative wood; bottom material mixed sand and gravel (50:50) with diameter of up to ⅜ inch (8 mm).	pH about or slightly over 7, dH 12° or more.	Two to three pairs of Central American cichlids, such as: *Thorichthys meeki* or another *Thorichthys*, "*Cichlasoma sajica*," *Herotilapia multispinosa*, or one pair of large Central American cichlids, such as *Paratheraps* (formerly *Cichlasoma*) *synspilum* or some other similar species.
4. Large community tank without plants for Central American cichlids. (This tank type also is suitable for South American cichlids such as *Uaru amphiacan-thoides* that attack plants.) At least 52 inches (130 cm)	Same as under 3.	Same as under 3.	Three to four pairs of Central American cichlids (see aquarium type 3); depending on tank size, more pairs, or two to three pairs largish Central American cichlids, such as *Herichthys cyanoguttatum* or "*Cichlasoma*" (*nigrofasciatum*). Best combined with large mailed catfishes. If tank is filled with brackish water, it also is suitable for *Etroplus suratensis*.

Aquarium type/size	Interior	Water	Occupants
5. Malawi Lake tank for Mbunas. At least 40 inches (100 cm)	Rock structures reaching to top of tank; lots of caves and hiding places. No wood! For the bottom, use finest available sand for foreground; farther back, in plant area, mixed sand and gravel (50:50) with diameter of up to ⅜ inch (8 mm).	pH definitely above 7, dH 12° or higher.	Up to three groups (one male and three to four females in each) of smallish *Pseudotropheus*, *Melanochromis* or *Labidochromis*. Example: 1,4* *P. lanisticola*, 1,4 *M. johanni*, and 1,4 *Labeotropheus trevawasae*, or 1,3 *P. zebra* and 1,3 *L. fuelleborni* (other combinations are possible). Longer tanks can accommodate four to six different groups.
6. Lake Malawi tank for *Aulonocara* or *Haplochromis*-type cichlids. 40 inches (100 cm)	Rock structures reaching to top of tank; lots of caves and hiding places. No wood! Hardy plants standing alone (*Anubias*) or in clusters (*Vallisnaria*); bottom should consist of fine, light sand.	pH around or clearly above 7, dH 12 or higher.	Two groups of 1,4* *Aulonocara* or smallish *Haplochromis*-type cichlids, such as *Aulonocara nyassae* and *Copadichromis boadzulu*, or *Aulonocara baenschi* and *A. jacobfreibergi* (many other combinations are possible). Can be kept together with largish catfishes (*Synodontis*, such as *S. multipunctatus*, or mailed catfishes).
7. Same as 6. but over 50 inches (130 cm) long.			In larger tanks four or more groups can be kept.
8. a) Lake Tanganyika community tank, or b) Lake Tanganyika tank for mouthbrooders that feed on aufwuchs, or c) Lake Tanganyika tank for secret brooders. 48 inches (120 cm) or more.	Rock structures reaching to top of aquarium; many caves and hiding places. No wood! Hardy plants standing alone (*Anubias*) or in clusters (*Vallisnaria*, including giant eelgrass); Bottom: fine, light sand in foreground; farther back, mixed sand and gravel (50:50) with diameter of up to ⅜ inch (8 mm) in planted area.	Same as under 7.	a) One group of *Tropheus moorii* or *T. duboisi* (1,4*) and one pair bottom-dwelling cichlids and four to five small *Julidochromis*, *Telmatochromis* or *Neolamprologus* (many combinations possible). b) Two groups *Tropheus moorii* and/or *T. duboisi* (1,4*; select only varieties of very different appearance), one pair *Eretmodus cyanostictus*. c) Two to three pairs smallish secret brooders, for example: *Neolamprologus longior* and *Julidochromis marlieri* and *Telmatochromis bifrenatus*.
9. Lake Tanganyika tank for cichlids using snail shells. 24 to 40 inches (60–100 cm)	Cover bottom with about 3 inches (8 cm) fine sand; rock structures in the back or plants along walls; snail shells (at least one per fish).	pH definitely above 7; dH 12° or higher.	A sizable group of *Neolamprologus multifasciatus* (two males and up to ten females), or one to three pairs of another *Neolamprogus* species.

*the numbers are a code indicating the ratio between the sexes; the first number indicates males, the second, females.

Water: 4–15° dH; pH 6–7; 72 to 82°F (22–28°C).

Companion fish for: South American and West African dwarf cichlids.

Live-bearing Tooth Carps

Genera/species: The larger (over 2 inches [5 cm]), natural or selectively bred varieties of the genera *Peocilia* and *Xiphophorus*, such as mollies, swordtails, and platys.

Life pattern: These fishes prefer the central and upper regions of the aquarium. They have interesting social and reproductive behavior patterns (good opportunity for observation!).

Feeding: Live, frozen, and dry food; plant matter is an important part of the diet.

Water: 10–25° dH, possibly with addition of salt; pH 7–8.5; 72° to 82°F (22–28°C).

Companion fish for: The smaller of the Central American cichlids, as well as for *Etroplus maculatus*. Though these fishes come from different habitats, they can be combined with dwarf cichlids from East African lakes. As long as the water quality is adequate, they also can be kept with other cichlids.

Rainbow Fishes

Genera/species: *Melanotaenia, Chilatherina*, and *Glossolepis* species.

Life pattern: Shoal fishes (keep at least six) for tanks at least 40 inches (100 cm) long. They prefer the upper reaches of the water.

Feeding: Live, frozen, and dry food.

Water: 10–20° dH; pH 7–8; 72° to 82°F (22–28°C).

Companion fish for: All small and medium-sized cichlids, as long as the requirements for water quality are met.

Catfishes

Genera/species: Armored catfishes of the genera *Corydoras* and *Brochis*, antenna catfishes (*Ancistrus* and similar species), mailed catfishes (*Hypostomus, Pterogoplichthys, Panaque*), and African catfishes of the genus *Synodontis*.

Life pattern: Armored catfishes live in groups, the others are solitary and bottom dwelling or in constant close proximity to some substrate (such as wood, rocks, bottom). Energetic and hardy.

Feeding: Omnivorous; favor worms and pellets made of plant matter; antenna and mailed catfishes like to graze on algae.

Water: 5–20° dH; pH 6–7.5; 72° to 82°F (22–28°C).

Companion fish for: All kinds of cichlids.

The Right Aquarium for Cichlids

Most cichlids have very similar requirements. The same kind of tank, interior setup, and planting are appropriate for most species. The same applies to the chemical properties of the water. This chapter will tell you what to watch out for when setting up an aquarium in which cichlids will thrive.

Safety Precautions

A number of electrical devices (lamps, heater, filter) are required to run an aquarium. Whenever water is close to electricity, there is potential danger. In order to avoid accidents, you should be sure to observe the following safety rules:

- All electrical appliances in or near the aquarium, especially lamps and tank covers, should be UL approved.
- Electrical equipment that will be operated inside the aquarium should carry a label saying that it is designed for such use.
- Unplug all electric wires before you handle anything in the water.
- It is a good idea to install an electronic safety device (available at aquarium stores or from electricians' suppliers) that will shut off the current if any apparatus or wire malfunctions.

What Kind of Aquarium Is Best?

Material: For cichlids, frameless glass tanks caulked with silicone rubber are best. They are available in many sizes and formats and even can be made to order if desired. Tanks with frames made of plastic, stainless steel, or anodized aluminum come only in standard formats, which are not always appropriate for cichlids. Plastic tanks are made only up to a certain size and therefore should be used only as temporary housing for cichlids.

Size: An aquarium for cichlids should measure at least 40 inches (100 cm) in length. Tanks longer than 60 inches (150 cm) are required only by very few species, such as *Paratheraps synspilum.*

Roots as aquarium decoration. In tanks decorated with wood—especially bog wood—only cichlids that prefer soft, slightly acid water should be kept.

A rough rule for calculating the amount of space needed by aquarium fish is to figure 1.5 to 2 quarts (1.5–2 l) of water per 3/8 inch (1 cm) of fish length (based on adult size). The amount you end up with stands for the water needed and doesn't include the space taken up by bottom material and decorations.

Important note: If you plan to set up an aquarium, ask an architect (sometimes the landlord can tell you) how much weight the floors of your apartment are designed to support. One liter of water weighs 1 kilogram. Therefore an aquarium measuring 40 by 20 by 36 inches (100 × 50 × 40 cm) weighs 440 pounds (200 kg), because it holds 53 gallons (200 l). People living on upper floors or in old buildings especially have to be careful about checking the strength of their floors.

Format: Cichlids are bottom-dwelling fish, and a tank designed for them should have as large a bottom area as possible to provide them with conditions close to those of nature. The tank also should be at least 16 inches or, better yet, 20 inches (40 or 50 cm) deep.

The following are typical standard measurements for cichlid tanks: 40 by 20 by 20 or 52 by 24

Aquariums with rocks. A tank with tall rock structures is especially well suited for cichlids from lakes Tanganyika and Malawi.

by 20 inches (100 × 50 × 50 or 130 × 60 × 50 cm). Only the high-backed swordtails and discus fish need tanks that are taller than 20 inches (50 cm).

Cover: Cichlids, especially during aggressive encounters, tend to jump out of the aquarium. Be sure, therefore, to cover the tank tightly with glass.

Location: Because daylight encourages algae growth, you should place the tank as far from windows as you can. Having a water faucet and a drain close by is a great convenience (see Changing the Water, page 21), and if the tank is located near a comfortable seat, you probably will spend more time watching the fish.

Heating

Most cichlids require water temperatures between 75° and 81° F (24–27° C), which can be maintained only with the help of a heater. Thermostatically controlled heaters, sold by aquarium stores in various designs, work best for this purpose. Ideally they are installed in a separate chamber of the interior filter (see next column). In any case, the heater should be sufficiently out of the way that a large cichlid with excess energy to work off cannot damage it. Heating cables on the aquarium floor and flow-through heaters are not a good idea for cichlids. Most of these fish burrow in the ground at least occasionally and would expose the heating cables. Flow-through heaters can be installed only in combination with enclosed external filters, and I can recommend their use only with major qualifications.

Filtering

With very few exceptions cichlids are voracious eaters, and it is therefore not surprising that large amounts of feces tend to accumulate in their tanks. To keep the water clean, a highly effective filter is necessary.

Internal filters with multiple chambers are especially practical for this situation. If you like to do things yourself and are handy, you can mount this type of filter yourself with some silicone rubber, or, if you prefer, you can ask someone at the aquarium store to do it for you. A centrifugal pump works best for operating this kind of filter. I recommend coarse, green filter wadding (available at aquarium stores) as the main filtering material. Only a thin layer of fine wadding is needed, 3/8 to 5/8 inch (10–15 mm). It functions as a preliminary filter and should be changed every week. Other types of filtering materials (tiny clay tubes, crushed lava) are also good filter fillers, but they are not very practical because they are difficult to remove and clean.

Internal foam filters, which are attached to the tank with suction cups, are in my opinion suitable only for tanks up to 32 inches (80 cm) long. Their filtering action is too weak for bigger tanks, or else such large models have to be used that they

Chromidotilapia guentheri from West Africa. This species is a pair-forming mouthbrooder. Immediately after the female releases the eggs (below), the male (above) picks them up in his mouth. ▶

cannot be hidden behind the decorations. However, in small tanks used for rearing fry they are ideal.

Fast filters with large filter inserts are, in my opinion, excellent. But they must be cleaned regularly. Because of their great mechanical filtering capacity they remove a lot of debris from an aquarium. Fast filters are attached by means of suction cups to the back wall close to the top.

External filters, either enclosed or open, are not suitable for tanks with cichlids because they don't filter enough and also are hard to clean. If you want to use an external filter, select a very large one.

Lighting

Cichlids come from waters with vastly different light conditions. However, hardly any of them seem unhappy if the lighting in their tank doesn't correspond to what they are used to in nature. So you can't go very far wrong. The main thing is to provide enough light for the plants in the aquarium. And the lighting should, of course, bring out the full color of the fish.

The length of artificial lighting is best regulated with an automatic timer and should be on about 12 to 14 hours a day.

Fluorescent light tubes are the best form of light for cichlid tanks. Most commercial aquarium hoods have fixtures designed for them. Fluorescent light tubes come in different shades of light. For cichlids I recommend using "daylight" or "white-tone," either in combination or alone. But the intensity of fluorescent lights deteriorates rapidly, and if fluorescent tubes are used about 14 hours daily, they lose half of their power after about six months. Replace them at least once a year. If there are several tubes, change one tube at a time in a rotation system to minimize the fluctuation of light intensity. If there are no plants in your cichlid aquarium, you don't have to be so conscientious and can wait considerably longer to change the tubes. Fluorescent tubes with a lot of red light should be avoided because they distort the natural colors of the fish. Red light also encourages algae growth.

My tip: Two fluorescent tubes are standard for a tank measuring about 40 inches (100 cm); if the tank is more than 52 inches (130 cm) long, use three tubes.

Mercury vapor lights and halogen metal vapor lights especially are good light sources for tall aquariums with lots of plants, but are not common in the United States. If you can find them, they can be placed on the aquarium or suspended above it. They are considerably more expensive than ordinary fluorescent tubes. Because most cichlid tanks are not very tall and therefore don't require great light intensity, aquarists generally choose the cheaper light source.

My tip: You can create the kind of muted light many of the shyer cichlids are comfortable in by placing floating plants in the aquarium to shade the water (see Aquarium plants, page 23).

Bottom Material

River gravel, mixed with sand, makes an excellent bottom material or substrate for most kinds of cichlids. Use half gravel with a diameter of 1/8 to 3/8 inch (3–8 mm) and half sand with a diameter of less than .04 inch (1 mm) (available at aquarium stores or at building supply stores).

Pure sand (the finest grade you can find) is needed for cichlids that chew through the ground, as do the *Geophagus* cichlids, and for ones that burrow a lot, like the *Neolamprologus* cichlids from Lake Tanganyika that use empty snail shells.

◀ Cichlids from Lake Tanganyika.
Above left: Striped Goby Cichlid *(Eretmodus cyanosticus)*; above right: *Tropheus duboisi;* middle: *Neolamprologus spec. "daffodil";* below left: Marlier's Julie *(Julidochromis marlieri)*; below right: *Neolamprologus brevis.*

Important: Rinse gravel and sand to be used in an aquarium long enough for the water to run absolutely clear. This way all the fine dirt particles in the bottom material are washed out.

Unsuitable for cichlid tanks are crushed basalt and lava because the edges of the rock grains are too sharp. Nor can peat or peat pellets be used because the peat acidifies the water and constantly would be stirred up by the cichlids.

Planting

Select only plants with largish, tough leaves for a cichlid tank. Amazon sword plants of the genus *Echinodorus,* African *Anubias* or Asian *Cryptocoryne* plants, Java fern, and various *Vallisneria* species are good choices (see Aquarium Plants, page 23).

All plants in a cichlid tank should be well anchored down. Place largish pebbles around the bottom, that is, the root, of each plant, so that the cichlids can't dig into the bottom there.

Note: In the case of many larger cichlids from Central America and Africa (such as *Paratheraps synspilum* and *Tilapia mariae*), there is no point in planting the tank because the plants are pulled or dug up or eaten despite all the precautions taken.

Decorative Materials

Rocks are essential for most cichlid species. Arrange the rocks so that they make caves of various shapes and sizes (see drawing on page 16). These caves serve as hiding places and are also potential spawning and brooding places. How many rock clusters you should set up depends both on the size of the tank and on the habits of the cichlid species you want to keep. For open-water spawners (see page 40) like *Anomalochromis thomasi* and *Laetacara curviceps*, caves are not as crucial as, for instance, for secret brooders (see page 40) of the genera *Apistogramma* and *Julidochromis*.

Flat rocks, such as pieces of slate, set up against the tank walls look very attractive and also have the advantage that fish can hide behind them when under attack from others.

Important: Make sure that none of the rocks you place in a tank has any shiny, metallic flecks because some metals give off toxic substances.

Wood from bogs (available at aquarium stores), especially if it is well soaked, is very useful for constructing hiding places and shelters for South American and West African cichlids.

Caution: Wood from bogs increases the acidity of the water and is therefore not suitable for aquariums in which cichlid species from lakes Tanganyika and Malawi are kept, such as *Pseudotropheus zebra* or *Julidochromus marlieri.*

Coconut shell halves (see drawing, page 44) from which all the loose fibers have been scrubbed away with a wire brush make excellent caves and shelters, as do clay flower pots. In both cases, an entry hole appropriate for the size of the fish has to be sawed or chipped out.

The Water

One of the most important conditions for keeping healthy fish, is, of course, that the water in the tank has the right qualities.

The right tank water for cichlids has a pH value of 6.5 to 7.5. (pH, which is the standard unit for measuring the acidity or alkalinity of water, stands for "potential of hydrogen"). The correct hardness lies between 8 and 15° dH. (Degrees of dH indicate the German system of measuring water hardness, where the letters stand for "deutsche Gesamtharte" [German combined hardness]. This "combined" hardness is the sum of carbonate and noncarbonate salts of calcium and magnesium present in the water.) Water with the indicated acidity and hardness levels is appropriate for cichlids. In many places, ordinary tap water corresponds to these values.

Exceptions to this general rule are a few South American and West African cichlids, such as the rarer species of the genera *Pelvicachromis* and *Apistogramma*. For these fish, I recommend water that is acid and soft, having a pH of 6.5 to 5.5 and a hardness of 2 to 8° dH. To achieve this water quality the tap water usually has to be especially treated, as with an ion exchanger. For more information on this procedure, consult the bibliography on page 68.

Note: The simplest way to check water hardness and acidity is to buy easy-to-use test solutions available at pet or aquarium stores. The pH value also can be measured with test strips and sticks.

Biotope Aquariums

A biotope aquarium is a tank in which the hobbyist attempts to recreate the characteristic habitat of one or several cichlid species as accurately as possible. In the previous paragraphs I have described how a basic aquarium for cichlids can be set up. The table on pages 12 and 13 contains more information and gives examples of biotope aquariums that meet the needs of cichlids with more specialized requirements.

The Quarantine Tank

Newly purchased cichlids always should be quarantined for three to four weeks before they join other fish. Even if they show no visible signs of illness, the fish may have parasites. Set up a small aquarium without bottom material or plants, but provide some hiding places by adding some flower pots or flat rocks (see Decorative Materials, page 20). Install a small internal foam filter to keep the water clean. In this setting you will be able to watch the cichlids well and nurse them back to health if a disease manifests itself.

Regular Maintenance

Maintaining an aquarium properly results in an optimal environment for your cichlids and ensures conditions that will allow your fish to stay healthy. The following chores are required:

- Change the tank water regularly.
- Clean the filter regularly.
- Remove dead fish and dying plant parts as soon as you notice them.
- Siphon detritus off the bottom with a glass tube of from 1/4 to 1/2 inch (6–12 mm) in diameter or a glass-rubber device (available in pet stores) every time you change the water.
- Always remove algae on the tank walls and the cover with filter wadding before you change the water.
- Collect snails.

Caution: Don't manipulate things in the aquarium more than necessary. Cichlids, more than other types of tropical fish, have a tendency to start fighting when bothered. Rearranging plants or decorative items affects the territories the fish have established and often gives rise to violent conflict.

Changing the Water

Changing the water is important especially in the case of cichlids because they are voracious eaters and quickly contaminate the water with feces.

In both community and species tanks the water therefore should be changed weekly, or, at the very least, biweekly. Replace one quarter of the tank water with fresh water; in the case of large cichlids or exceptionally voracious ones, up to as much as two thirds of the water should be changed at a time.

In a tank containing mostly dwarf cichlids, only about one third of the water needs to be replaced every two weeks.

Just how you go about changing the water depends primarily on the size of the tank. You can use a bucket and a hose about 6 feet (1.5 m) long to change the water of a smallish tank. (Never use the bucket for any other purpose so that the fish will not accidentally have contact with cleansers or similar substances.) With larger tanks, a hose running from the tap to the tank and another one leading to a sink can be used.

Check ?

My tip: If you use tap water and a hose, direct the water into the tank in a sharp spray or run it through a percolator. This way a large portion of the chlorine present in tap water is dissipated, and fresh oxygen is introduced into the tank water.

With most cichlids it is not necessary to prepare the new water in any special way. All you need to do is to make sure the new water is the same temperature as the "old" tank water. This you can do by turning the hot water faucet on just enough or, if you don't have a mixing faucet, by pouring warm and cold water in a bucket and checking it with a bath thermometer. If you have cichlids that need acidic water (see the descriptions starting on page 49 and Breeding Cichlids, page 40), you can condition the tank water by filtering it through peat.

Note: If the water temperature in the tank drops about 5° F (2–3° C) as a consequence of the water change but the heater raises it again to the proper level during the next 24 hours, this in no way harms the fish; on the contrary, it has a healthy, hardening effect on them.

Cleaning the Filter

The larger a filter is in relation to the amount of water in the tank, the less often it has to be cleaned. I cannot give you an exact cleaning schedule here. But don't clean your filters too early (except for filter inserts) because they only start filtering really well once they have been in use for a while. Even though at that point they work more slowly, they filter more thoroughly because biological filtering action has started.

Note: Don't use hot water or chemicals to clean filters. Hot water and chemicals destroy the bacterial flora on the filter substrate, and bacteria are important because they contribute to the filtering effectiveness.

Internal filters with several chambers should be cleaned when the fine filter wadding in the preliminary filter is getting really dirty. Replace or wash this layer of wadding. The coarser filtering material (coarser wadding, crushed lava, tiny clay tubes) in the main section of the filter needs to be cleaned only at longer intervals (every three or six months or, under certain conditions, after more than a year) and never should be replaced. This way the biological filtering capacity of the material is not interfered with. To clean this part of the filter, remove the filtering mass, rinse it, siphon off the debris from the bottom of the chamber with a hose, then replace the filtering material.

Internal foam filters used in small rearing tanks are cleaned by rinsing and squeezing the foam under running cold water until no more dirt emerges.

Filter inserts are taken out of the tank and the material in them either replaced (fine filter wadding) or washed (foam cartridges, coarse wadding) as described above.

Enclosed external filters can be cleaned as follows. Disconnect the filter from the hose, open it, remove the filtering material, and then clean or replace this material as described above.

Important: Don't change the water and clean the filter at the same time! This would endanger the health of the fish.

Filter hoses and pipes as well as pump housings should be cleaned every three to six months. This can be done with special hose brushes (available at aquarium stores) or with some filter wadding that is pushed through the hoses or pipes with a long, thin rod.

Plant Care

In the case of most cichlid tanks, there is little need for plant care. Where plants are used, taking care of them involves the usual jobs, such as picking off and removing old leaves, shortening long-stemmed plants, and thinning.

The best time to fertilize the plants is immediately after changing the water and perhaps once more between water changes. Aquarium stores sell fertilizers for water plants suitable for use in fish tanks.

Caution: Never use fertilizers designed for house plants! These are deadly for fish.

Aquarium Plants

Anubias barteri var. nana

Origin: West Africa.

Care: Robust, undemanding; slow growth; transplant as little as possible.

Light: Dim to moderate.

Water: Up to 15° dH; 72° to 82°F (22–28°C); pH 6.0–7.5.

Propagation: Side shoots from root stock.

Location: Anywhere in the tank, on top of rocks or roots.

Ceratophyllum demersum

Common hornwort

Origin: Found worldwide.

Care: Floating water plant, fast growing; should be thinned periodically.

Light: Undemanding, thrives even in moderate light.

Water: Up to 20° dH; 50° to 82°F (10–28° C) but best below 75°F (24°C); pH 6.0–7.5.

Propagation: Side shoots.

Location: Anywhere in the tank; is not bothered even by plant-eating cichlids because they usually don't like the tough leaves.

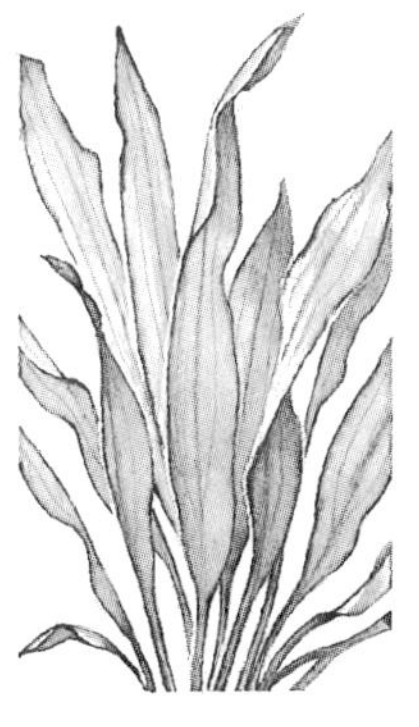

Echinodorus amazonicus

Amazon sword plant

Origin: Brazil.

Care: For tanks 40 inches (100 cm) or longer; fine, loose bottom material.

Light: Medium bright.

Water: Up to 10°dH; 75° to 82°F (24–28°C); pH 6.5–7.5.

Propagation: Adventitious plants form on flower stalks.

Location: Plant singly in the middle area of the tank; this plant needs a lot of room.

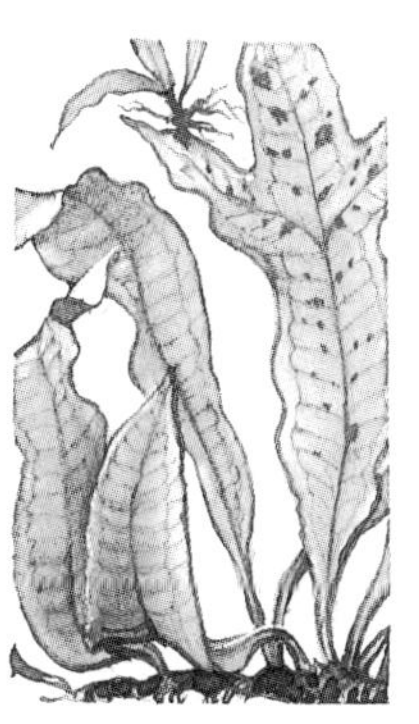

Microsorium pteropus

Java fern

Origin: Southeast Asia.

Care: Attach with nylon thread to roots or rocks, where it will root.

Light: Dim to moderate.

Water: Up to 15°dH; 68° to 82°F (20–28°C); pH 6–7.5.

Propagation: Adventitious plants grow on leaves; splitting of root stock.

Location: Anywhere in the tank except the foreground.

Ceratopteris thalictroides

Water sprite

Origin: Africa, America, Asia, northern Australia.

Care: A prolific floating plant that also can be set in shallow ground; the top of the root must remain exposed; needs a lot of light.

Light: Bright; use at least two fluorescent tubes.

Water: Up to 15°dH; 75° to 82°F (24–28°C); pH 6.5–7.5.

Propagation: Adventitious plants form on leaves.

Location: Middle of the tank or along walls.

Vallisneria spiralis

Eelgrass

Origin: Tropical and subtropical zones anywhere.

Care: Undemanding; top of root must not be buried in substrate.

Light: Medium to bright.

Water: Up to 20°dH; 59° to 86°F (15–30°C); pH 6.5–7.5.

Propagation: Runners.

Location: Background and along sides of tank; also clusters in the middle.

The Right Aquarium for Cichlids

Reconditioning an Aquarium

If an aquarium containing cichlids is properly maintained, it should not need reconditioning more often than every two or three years. Reasons for reconditioning are:

- bottom material getting encrusted and
- constantly cloudy water.

When you do have to recondition an aquarium, place the fish in another tank for a few days. Use well-filtered water that has been left standing for some time to fill this tank, and add some rocks to supply hiding places for the fish.

Rinse the bottom material of the "old" tank or replace it with fresh material. Then arrange things inside the aquarium again, and fill it with fresh water.

Note: Never recondition a tank unnecessarily. This procedure puts the fish under considerable strain and can upset or destroy the pair bonds between cichlids.

Setting up a Fish Tank Step by Step

1.

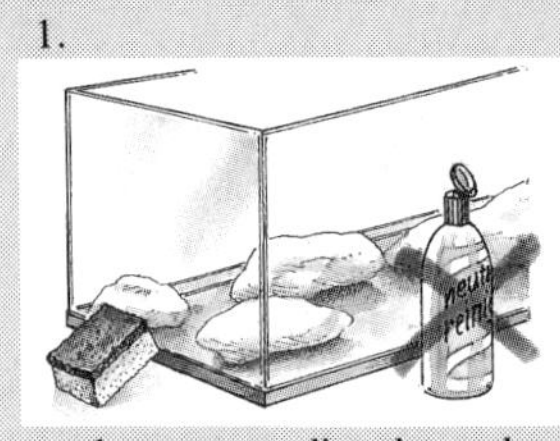

Wash the tank with lukewarm water, using a new plastic pot scrubber or a handful of filter wadding. Don't use any kind of cleanser! Place the bottom stones of the rock structures directly on the glass.

2.

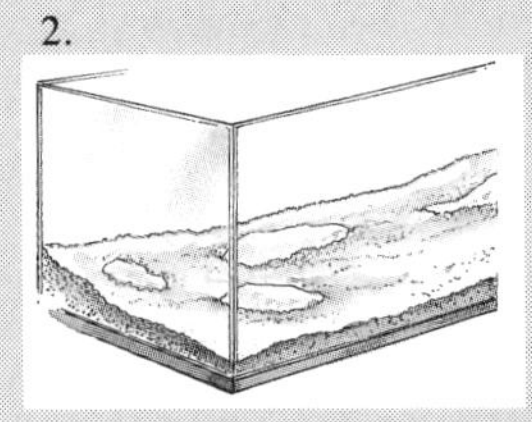

Rinse the bottom material (sand, gravel) thoroughly (see page 20) and put a layer two to four inches (5–10 cm) deep (rising toward the back) in the tank. If available, mix in a handful of bottom material from a functioning, healthy tank.

3.

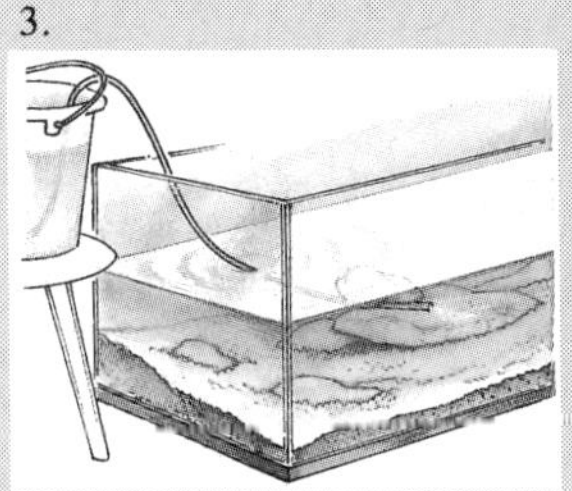

Fill the tank about half full. Let the water run in over a flat rock so that it will not stir up the bottom.

4.

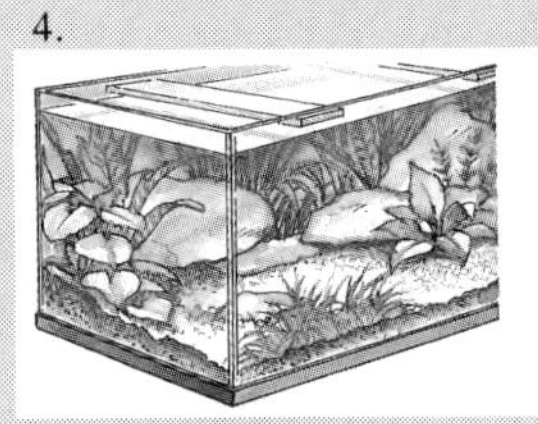

Complete the rock structures. If there are to be plants, plant them now along the back and sides of the tank. Add the rest of the water up to within 1/2 inch (1 cm) of the strips that are glued to the tank sides for the top to rest on.

5.

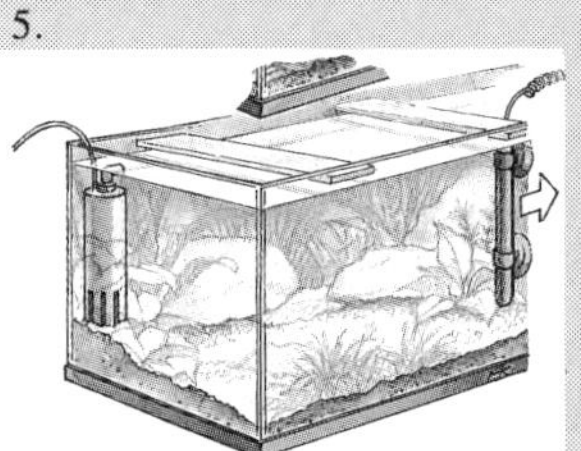

Install the technical apparatus (heater, filter) and, if possible, "inoculate" the filter and substrate with some material from a well-functioning filter in another tank.

6.

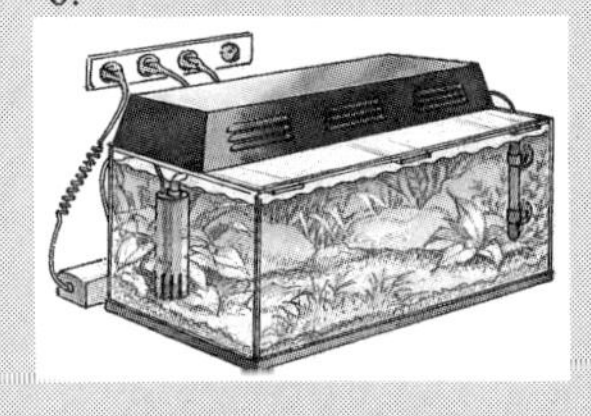

Put the tightly fitting top in place, and install the lights. Turn on all the technical accessories; then wait three days or, in a planted aquarium, a week before introducing the fish.

Proper Nutrition

What You Should Know about the Food Habits of Cichlids

Basically, feeding most kinds of cichlids doesn't present any great problems. Almost all cichlids will get used to fish food that is obtainable easily by the hobbyist, even if the fish have in nature specialized in certain foods. (The descriptions of individual species, starting on page 51, provide feeding suggestions.)

The great majority of cichlids are *omnivores*. In nature, as well as in an aquarium, these fish feed mainly on live food like mosquito larvae, tiny crustaceans, and worms, but, when hungry, they also will nibble on dead fish, eat plants, burrow through the debris at the bottom in search of food, and even go after smaller fish. Omnivorous fish living in an aquarium need a nutritious and varied diet (see feeding rules, page 32).

There are also cichlids that are *pure carnivores*. They have specialized in hunting other fish, which have become their exclusive diet. The *Crenicichla* species of South America are such carnivores, as are some of the larger "*Haplochromis*" species from Lake Malawi in Africa. It might be assumed that these fishes cannot be kept in a tank because of their feeding habits, but this is not the case. "Hunters" can, with a few exceptions, learn to accept "dead" food, such as beef heart or frozen fish filets (see Frozen Live Food, page 30). The main thing to be careful about is what other fishes they are combined with (see page 10). These other fishes should be of comparable size so that they will not be regarded as a potential meal by the meat eaters.

Plant-eating cichlids are not all the same and have to be treated accordingly. There is a difference between fish that feed on plants only when there is nothing else to eat and fish that like the taste of plants or—as some do—eat mostly plants. The first group will not bother aquarium plants at all if given a varied diet, whereas the second will wreak havoc in the greenery. Among the latter are cichlids that grow quite large, such as *Uaru amphiacanthoides* and many *Tilapia* species.

Finally there are so-called algae or aufwuchs feeders (see drawing, page 30) among the plant-eating cichlids. (Aufwuchs is a term describing the organisms that live on the surface of submerged objects like plant parts, stones, and rocks but do not put out roots or otherwise penetrate into them.) The cichlids feeding on aufwuchs live mostly on the algae that grow in a thick layer on the rocks in the water. In the aufwuchs layer there are many animal organisms, which the cichlids eat along with the algae. Aufwuchs feeders most commonly kept by hobbyists include the *Tropheus* cichlids of Lake Tanganyika and the Mbuna cichlids of Lake Malawi that belong to the genera *Pseudotropheus*, *Melanochromis*, and *Labeotropheus*.

Vegetarian fish food is discussed on page 31.

Note: The digestive system of *Tropheus* cichlids is especially adapted to vegetarian food, and these fishes should therefore not be combined with carnivorous ones.

The Different Kinds of Fish Food

In this day and age feeding aquarium fish has become very easy, for pet and aquarium stores now sell fish food in all conceivable forms and compositions. There is dry food, live food, and frozen food—the choice is vast. Aquarists who don't want to rely solely on what is available at shops also can prepare food for their cichlids themselves and store supplies in the freezer.

Dry Food

Dry food can be used as the basic staple for cichlids because it contains many important nutrients, including vitamins, trace elements, and roughage. However, there are some cichlids that adapt to eating dry food only after a hunger strike that can last several days. But once they become used to dry food, the fish thrive on it. Dry food comes in the form of flakes, pellets, and granules.

Food flakes are sold in different sizes. Always buy the larger flakes. These work fine as they are for medium size and large cichlids. For the smaller species, you will have to crumble the flakes.

There are two kinds of *food pellets*. One kind is stuck to the front wall of the tank. With this method of feeding, it is easy to check how much actually is eaten. The other kind of pellet simply is dropped into the water, and the fish either eat it as it sinks through the water or pick it up on the bottom. The pellets also work extremely well for small fish and even the tiniest fry. Rub the pellets, one at a time, between your fingers over the water surface. The tiny food particles—almost as fine as dust—which make up the pellet, first float on the surface and then drift slowly toward the bottom. This feeding method is especially appropriate for fry that like to stay in the middle layer of the water (see the descriptions starting on page 49). But feed only as many pellets as will be consumed within ten minutes. If too many are given, the water quality suffers, and the tank may become cloudy.

Food granules for tropical fish are very similar to food flakes. Here, too, you should choose the larger size. Food granules for trout or pellets for raising other food fish also fall into this category. Sometimes cichlids seem to like this kind of food, but it should not be given to them as an exclusive diet because it can lead to health problems.

Live Food

What is called "live food" really includes three different forms of food, namely, live food animals, frozen food animals (frozen food), and freeze-dried live food. All three are excellent for supplementing the diet of cichlids fed mostly with dry food.

You can buy live food animals, catch them yourself, or raise them (see page 31). Buying them saves you time, and you can be relatively sure that you are not inadvertently introducing pathogens or parasites into your tank. If you collect food animals yourself, be sure they come from clean waters where there are no fish. Live food from fish ponds can harbor pathogens or parasites, such as fish leeches and fish lice.

Caution: In some places, state and federal environmental protection legislation prohibits removing flora or fauna from public waters. Find out what regulations apply to the species in question by inquiring at appropriate authorities. For privately owned bodies of water, you should, of course, obtain the landowner's permission.

Suitable Food Animals

I have fed all the food animals discussed here to my cichlids and have found that most species eat them eagerly.

Red mosquito larvae live in the bottom mud of waters with high levels of organic matter. Depending on where they come from, the larvae may contain substances (such as heavy metals) that are toxic for cichlids. It therefore is important to obtain these animals from a reliable source. Catching red mosquito larvae yourself is difficult, but they are regularly sold at pet stores. Don't buy more than your fish will eat within two days, or freeze the rest to have on hand later.

Black mosquito larvae (biting mosquito larvae) are an excellect food for cichlids. Many breeders even claim that feeding these larvae to breeding stock improves spawning. These larvae are found in small accumulations of water directly under the surface from early spring to late fall. If you decide to collect some, take only as many as

One of the best known cichlids. The Butterfly Dwarf ▶ Cichlid or Ram *(Papiliochromis ramirezi)* from South America. The fish pictured here is a female.

you can feed your fish within a day or two. Store the larvae in a jar of cold water with a cover. When ready to remove some of the larvae, briefly invert the jar over the bathtub before taking off the lid. This way the larvae that have hatched drown. Some hobbyists, who don't want to take the risk of releasing mosquitoes into their homes, make a point of freezing the larvae before feeding them to the fish.

White mosquito larvae (phantom midge larvae) are almost as transparent as glass. They hover horizontally in clear, still water. You can catch them only rarely, but they are sold live at pet stores. These larvae are a truly excellent food for cichlids, but the small amounts you get when you buy them rarely are enough for more than one feeding.

Water fleas, including *Daphnia* and *Cyclops*, form an important part in the diet especially of small cichlid species and of fry. Water fleas contain a lot of roughage, as well as minerals and vitamins. How beneficial the nutrients supplied by these minute crustaceans are is shown by the exceptionally bright coloring a number of cichlids (such as *Neolamprologus leleupi* and *Melanochromis brevis*) exhibit when they are fed water fleas regularly. Water fleas live in clean, still water.

Tubifex worms are the best known of the worms used to feed tropical fish. They are almost always available at most pet stores. Cichlids like them and consume them in great quantities. However, tubifex worms harbor two dangers. These red worms live in the mud of dirty water and therefore have to be rinsed several times in cool tap water. The dead worms float to the surface and can be poured off. Soaking the worms is a good idea so that they will discharge the muddy contents of their intestines. To soak them, leave them

◀ Cichlids from Lake Malawi.
Above: A male *Melanochromis johannii;* below: a male Nyasa Peacock *(Auloncara nyassae)*.

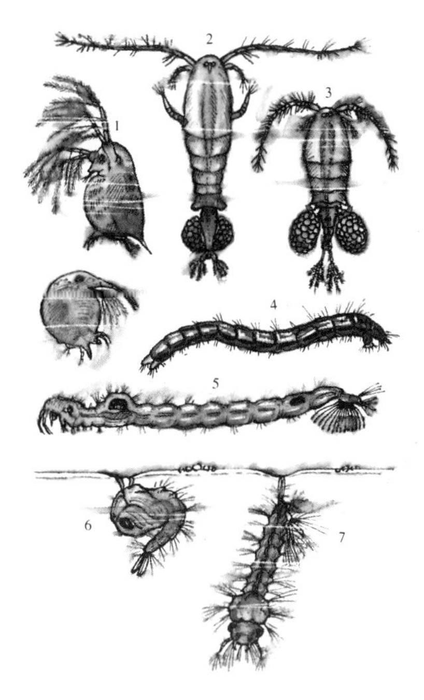

Food animals. Live food is an important component in the diet of your cichlids. You either can catch the food animals shown here yourself or buy them at pet stores. 1. Water flea *(Daphnia)*; 2. *Diaptomus* female with eggs; 3. *Cyclops* female with eggs; 4. Red mosquito larva *(Chironomos)*; 5. White mosquito larva *(Corethra)*; 6. Black mosquito *(Culex)* pupa; and 7. larva.

for several days in a small container with water dripping or running slowly into it from a faucet. You also can use a so-called tubifex spiral (available from aquarium stores or through mail-order firms) for this purpose. Tubifex worms treated this way are quite unlikely to be the source of disease.

There is one other potential danger in feeding tubifex worms to aquarium fish. Following their natural habits, worms that are not eaten right away try to escape into the bottom material, where they start to decay if they die. For this reason you should feed only as many worms as your cichlids will consume immediately. Another way is to put the worms in floating feeding funnels or in a shallow dish placed on the bottom of the tank.

Note: Tubifex contain practically no roughage, and cichlids often eat great amounts of them. This can give rise to intestinal upsets. You should therefore not feed tubifex worms to cichlids every day or give too many at a time.

Earthworms also make a good food for many cichlids of medium to large size. They have the advantage of being readily available in most places in considerable quantities.

Mealworms (the larvae of darkling beetles) and fly larvae (sold by suppliers for sports fishing) are also a good addition to the diet of large, meat-eating cichlids.

Frozen Live Food

The food value of frozen food is almost equal to that of live food animals, and frozen food offers the convenience of being easy to store and use. Pet and aquarium stores sell frozen food for tropical fish, or you can freeze food animals yourself when you catch a lot of them or can buy them at a good price. Live fish food is frozen just the same as food for human consumption. It retains its full nutritional value for one to two years.

Feed frozen food to fish only after it has thawed completely.

Food animals that lend themselves to freezing include all kinds of mosquito larvae, water fleas, and freshwater amphipods. Pet supply stores also sell brine shrimp (*Artemia salina*), opossum shrimp (*Mysis*), crabs, and sand shrimp.

Note: Frozen brine shrimp are salty and should be rinsed several times under running water before being fed to the fish. Give brine shrimp to cichlids only in rotation with other kinds of food.

Beef heart consists of very lean red meat and is an important food for cichlids that are highly carnivorous. Frozen beef heart is sold prepacked in various sizes at pet stores, or you can freeze beef heart yourself. Before you freeze it, remove all white tissue. Then run the meat through a meat grinder, grinding it coarser or finer according to the needs of your fish. Place individual serving portions into plastic bags and freeze in the usual manner.

You can buy *fish filets, mussel meat and shrimp meat* fresh and freeze them for your cichlids in individual portions, or these foods can be found already frozen in the frozen foods section of your supermarket. If you buy them already frozen, all you need to do is to break them up into individual serving portions.

Frozen food mixtures have been marketed by firms producing food for tropical fish for some time. They come in morsels of various sizes, so that you can pick the right size for different kinds of fish. Most often the main ingredient of these mixtures is beef heart, to which other things are added, such as vegetables like spinach, shrimp, or other crustaceans, to provide roughage, and, sometimes, vitamins. Unfortunately not all manufacturers list the contents on the labels. Some aquarists also make their own food mixtures for their fish, using primarily beef heart, fish filets, or mussels. I know fish fanciers who swear that adding grated carrot or paprika intensifies the colors of cichlids.

Algae-eating cichlid. Many cichlids from the lakes of the East African Rift feed on algae, which grow in thick layers on rocks and pebbles.

My tip: You can achieve the same variety of diet that food mixtures offer by rotating unmixed foods every day. In my opinion, this meets the needs of cichlids for a varied diet better than giving them the same food mixture day after day.

Freeze dried live food is just as good as any of the foods already described. Its only drawback is its relatively high cost, for cichlids go through impressive amounts of it. It is a decision every aquarist has to make for himself or herself.

Vegetarian Fish Food

Vegetarian food is high in fiber and therefore is an important element in the diet of cichlids. This is true not just for cichlids that feed on plants but for cichlids in general.

Some types of *dry food* contain higher amounts of plant matter than others and therefore are appropriate especially for cichlids. When you buy dry food, select the kinds that have a lot of plant matter. Often the names of the products will give an indication of what is in them, or else compare the list of ingredients on the labels.

Lettuce, preferably unsprayed or else very thoroughly washed, is a good food for plant-eating cichlids. Depending on the size and number of the fish, you can feed them a leaf at a time or submerge the central part of an entire head, weighing it down with a rock on the bottom of the tank.

Spinach leaves (again unsprayed or well washed) are used not just as a supplement to go along with a food mixture but can be fed as a meal by themselves.

Tender aquatic plants, such as waterweed (*Elodea*) and duck weed, are regarded by many cichlids as delicacies.

Food for Young Fry

Young fish need food with more protein than adult fish. The following kinds of food make a good rearing diet for fish fry.

Fry of Discus fish "grazing." Discus fish produce a special secretion on their bodies that the baby fish feed on when they start swimming freely.

Artemia nauplii, which are brine shrimp (*Artemia salina*) at their earliest stage of development, are usually the first food given to freshly hatched cichlids (for instructions on breeding Artemia nauplii, see page 32).

Dry food, designed especially for fish fry, is made by all major feed manufacturers and sold at pet and aquarium stores. Crushed food tablets (see page 26) also can be used as a rearing food.

Liquid food, in tubes like toothpaste, also is used for fry immediately after hatching. But the fry of very few cichlid species are so small that they require this type of food.

The microorganisms present in the sediment on the bottom and in the aufwuchs in tanks that have been in operation for some time make an excellent first food for newly hatched fry.

Raising Food Animals

Given the voracity of cichlids, it may be worth your while to raise the food animals that should be added to the basic diet for these fish yourself. Breeding colonies along with instructions, as well as breeding containers and special soil, are available through mail-order firms (check the classified sections of aquarists' magazines), at pet stores, or through aquarists' associations.

Various kinds of worms, such as white or grindal worms, microworms, and earthworms can be raised for cichlids. But they should be fed to cichlids only in combination with other types of food, for — with the exception of earthworms — worms are high in fat. If you fed your cichlids nothing but worms, they soon would develop serious health problems.

Artemia Nauplii

These larvae of brine shrimp not only are an excellent food for fry, they also are eaten enthusiastically by many smaller to medium size adult cichlids.

Hatching of the eggs: Rinse a 1-quart or liter bottle well and fill it about two-thirds full with water. Mix in 1 teaspoon of ordinary cooking salt (without additives) or sea salt for marine aquariums, and add half a teaspoon of Artemia eggs. Aerate the water in the bottle thoroughly. This can be done by attaching an extra air hose to your aeration pump and sticking the end of it into the bottle. Pet stores also sell special areation devices. Store the bottle at a temperature above 68°F (20°C).

Incubation period: 24–36 hours.

Removal: Attach a piece of aquarium hose to as large a syringe as you can get (from your veterinarian or at a drugstore). Let the bottle with the hatched larvae stand for a few minutes, then with the syringe siphon up some of the larvae that have settled on the bottom and give them to the baby fish. With fish that are particularly sensitive, you should rinse the nauplii under water in a special Artemia sieve (available from pet stores) before feeding them to the cichlids.

My tip: Small amounts of the nauplii can be hatched easily by pouring about 1 inch (2 cm) of properly salted water into shallow glass or plastic dishes and sprinkling some Artemia eggs over the water. The eggs take the same amount of time to hatch and can be removed as described above.

Ten Rules for Feeding Cichlids

1. Give different kinds of fish food in rotation to provide a varied diet, so that your cichlids will receive all the necessary nutrients in sufficient amounts.
2. Live food should be given only in small portions. Newly acquired fish and fish that haven't had any live food for some time especially are eager for it and may gobble it up at such a rate that they develop intestinal upsets that may in some cases be fatal.
3. The bigger the cichlids, the larger the food morsels that they can swallow. The size of the morsels plays a significant role in how much food individual species need to satisfy their hunger.
4. Baby fish should get only food that is no larger than their eyes. Increase the size of the morsels as the fry grow larger.
5. Feed fully grown cichlids twice a day.
6. Fry are fed four times a day. Feeding them this often is essential for quick and normal growth.
7. Give only as much food as the fish will eat up in about ten minutes.
8. Institute one day of fasting a week for fully grown fish.
9. If possible feed the fish once in the early afternoon and again in the evening, at the latest one hour before turning off the aquarium lights.
10. Turn off the filter when you feed the fish, so that the food is not sucked into the filter.

Diseases of Cichlids

Generally speaking, cichlids are subject to the same diseases that affect other tropical fish. Fish are particularly susceptible to disease if their organisms already are weakened by stress, or by being subjected to unfavorable living conditions. Unfortunately, treating sick fish is often quite problematical because only a few common diseases can be diagnosed definitely. When drugs are used to combat disease, they can have unanticipated side effects, as, in the case of cichlids, infertility. That is why the saying "prevention is better than any cure" is applicable particularly if you keep cichlids.

Note: If drugs do have to be resorted to, use only products that are designed especially for treating tropical fish and have been tested (available from pet stores or drugstores). Make sure you follow the directions to the letter.

Caution: Keep all medications out of reach of children!

Preventive Measures

Probably the most important thing you can do to discourage disease is to provide proper environmental conditions and optimal care for your fish. But there are a few other safety rules you should observe.

- Dead fish, decaying plant parts, and leftover food that is beginning to decompose should be removed from the tank immediately.
- Regard water from other fish tanks as a possible source of contamination, and prevent any contact of your fish with such water.
- Newly acquired fish, snails, and aquatic plants should, as a matter of principle, be quarantined for three to four weeks (see The Quarantine Tank, page 21).

The Importance of Disinfecting

New items for the tank (roots, stones, clay flower pots) should be rinsed thoroughly with very hot water. If these items have been used in another aquarium, place them first in the quarantine tank.

Fish nets and other accessories used in more than one tank should be rinsed with hot water before and after every use. Or you can have a separate net and set of tools for every tank. (But don't get them mixed up!)

A tank has to be cleaned and disinfected with special thoroughness if a number of the fish living in it have gotten sick and the tank is to be newly occupied. Use hot water to which you add table salt until it no longer dissolves. Wash the tank and let it stand with the salt solution in it for a day.

Note: Instead of salt and hot water, you can use disinfectants sold at pet stores for treating white spot disease (see next page). Prepare a solution as indicated in the directions for use, and let it stand in the tank for about a week.

Important: Rinse everything you have disinfected thoroughly with clear water.

My tip: The most important point is not to infect your healthy fish. If you have two or more tanks and some of them have been free of all disease for some time, don't introduce new fish into them, even after a quarantine period of three to four weeks. Wait several months during which you can observe the new fish in a separate tank; then combine them with your established healthy fish, but only if no disease has been detected during this time. If you follow this procedure consistently, all your tanks, except for the quarantine tank, should be disease-free. Of course you have to make sure diseases are not spread from the quarantine tank to other tanks on tools like fishnets.

Relatively Common Diseases

The following pages discuss some of the diseases most commonly encountered in cichlids. There are, of course, a great many others that possibly could turn up in your aquarium. (To find out more about fish diseases, see Addresses and Bibliography on page 68.)

Diseases of Cichlids

Note: Get in touch with experienced aquarists, breeders, or dealers when you first start out keeping fish. This can prove extremely useful especially if your cichlids get sick. Membership in an aquarists' club or the American Cichlid Association (see Addresses, page 68) is very helpful in establishing such contacts.

White Spot Disease or "Ich"

Signs of illness: Small dots up to 1/24 inch (1 mm) in diameter and looking like grains of salt appear on fins and skin.

Cause: Ciliate protozoans (*Ichthyophthirius*).

Treatment: Medications available from pet stores (follow directions). The treatment has to be repeated after 20 days.

Fungal Diseases

Signs of illness: Moldy growth on skin and fins, first in filaments, then forming white tufts like cotton wool.

Cause: Fungi of the genera *Saprolegnia* and *Achlya*. They develop most often on injured skin, as after a fight.

Treatment: Medications available from pet stores (follow directions). With East African and Central American cichlids, that is, fish that can take harder water, adding ordinary table salt (1 g per liter of tank water) or a medicinal salt preparation sometimes is effective in mild cases.

Fin Rot

Signs of illness: Frayed fins and tail. Progressive disintegration of tissue until, sometimes, only a stub of the tail remains.

Cause: Bacteria, especially of the genus *Pseudomonas*.

Treatment: Medications available from pet stores (follow directions). Careful monitoring of water quality and an occasional slight rise of the water temperature help prevent this disease.

Ascites and Tuberculosis

Signs of illness: Badly bloated body with protuberant scales, bulging eyes, lack of appetite, tumors.

Causes: Bacteria, viruses, fungi.

Treatment: There is no treatment that is likely to cure these conditions, and it therefore is better to kill afflicted fish (see 36 page). Try to prevent these diseases by providing optimal living conditions.

Hexamita Disease

Also commonly referred to as discus disease and "hole-in-the-head" disease.

Signs of illness: Lack of appetite; slimy, whitish feces; in some species, darkening of the body; gagging motions; and, at more advanced stages, holes in the head region with whitish growths in them.

Cause: Flagellate protozoans, such as *Hexamita* and *Spironucleus*.

Treatment: There are various medications available from your pet store. If the treatment is unsuccessful, this may be because other kinds of flagellate protozoans are causing the disease. Flagellates of different genera have to be combatted with different drugs. Hole-in-the-head disease, caused by *Hexamita*, often accompanies and aggravates tuberculosis.

Gill Flukes

Signs of illness: Accelerated breathing, reddening and swelling of the gills, rubbing the head or the gill covers along the bottom or against items of decoration. Gill worms are common in discus fish and angelfish but can affect other cichlids, too.

Cause: Gill flukes (*Dactylogyrus*, *Cichlidogyrus*).

Treatment: Difficult; pet stores sell various excellent medications to use for this condition. Be sure to follow the directions meticulously, and use

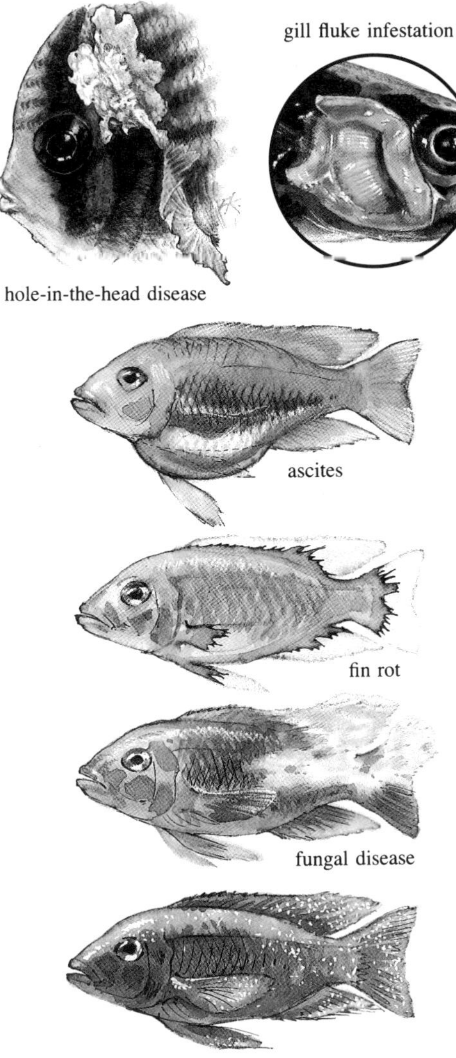

Diseases of cichlids. Shown here are the most common diseases affecting cichlids. They are caused by bacteria, viruses, and parasites.

only in water with a hardness of at least 15° dH. When used with fry or small cichlid species, the dosage has to be very exact and possibly somewhat reduced.

Nematodes (Threadworms and Roundworms)

Signs of illness: Emaciation of the fish, sometimes in spite of voracious appetite and sometimes as a result of refusing food. Occasionally worms protrude from the anus, looking, in the case of the thorny-headed worm (*Acanthocephala*), like the points of needles.

Cause: Threadworms, Roundworms.

Treatment: Pet stores and mail-order firms for some time now have been advertising (in aquarists' magazines) frozen red mosquito larvae to which a worming medicine has been added.

Poisoning

Signs of illness: Gasping for air at the water surface, accelerated breathing, draining of colors, swimming with clamped fins, and other abnormal behavior.

Causes: Poisoning can have very different causes, such as:

- buildup of products of organic decomposition, like ammonia, nitrite, and nitrate, if the water is not changed frequently enough (see Changing the Water, page 21) or if the tank is overcrowded;
- toxic chemicals, such as insecticide sprays used too close to the tank, or chemical cleansers;
- metals, such as copper and zinc, either pure or in chemical combinations (sometimes present in rocks with metal flecks; also in copper water pipes); and
- too much chlorine, which damages the mucous membranes and gills of cichlids.

Treatment: Move the fish immediately to uncontaminated water, or change some of the tank water.

Damage Due to Too Much Acidity or Alkalinity

Signs of illness: Damaged skin; formation of mucus; rapid, jerky movements.

Cause: Fluctuations in the pH value.

Treatment: Place fish in neutral water (pH 7) or water with the correct pH for the species in question (see descriptions starting on page 49).

Lack of Oxygen

Oxygen depletion not only affects the fish directly but also weakens them, so that they are more likely to succumb to diseases.

Signs of illness: Gasping for air at the surface, accelerated breathing, raising of gill covers.

Correction: Make sure that the tank is well aerated and the filter works properly; remedy the cause of oxygen depletion (excess food that is decomposing, dead fish hidden somewhere in the tank, broken down filter, and so on).

Effective Use of Medication

When medicating a tank it is important

- to provide good aeration,
- not to filter with activated charcoal, and
- to filter through charcoal afterwards in order to remove the medication from the water.

Note: The use of drugs can reduce the efficiency of a biological filter. This undesirable side effect cannot be avoided, but the beneficial bacteria will get reestablished in time.

Killing a Fish

Sometimes it makes sense to kill a fish with an incurable condition to spare it further suffering. If you have to do this, sever the fish's spine right behind the head with a sharp knife or with sharp scissors. With larger cichlids you should stun the fish first by hitting it on the head with a blunt object. This saves the fish from suffering unnecessary pain.

The Behavior of Cichlids

Surely the beauty of their colors is the prime reason why cichlids are so very popular with hobbyists. But the great variety of behavior of these fish must be a close second. There is hardly any other fish family that has evolved such a rich repertoire of behavior patterns as the cichlids. It is this same trait that has made cichlids a favorite subject of study for ethologists.

The "Language" of Cichlids

Cichlids have several means of conveying information to each other about their species, sex, and moods:

- coloration and color pattern;
- sound utterances;
- chemical signals; and
- complex behavior patterns.

Coloration and Color Pattern

Coloration, along with body shape, is what helps cichlids tell whether or not another fish belongs to their own kind. In many species the coloration differs between males and females and in this case signals gender as well. Coloration and color pattern determine to a large extent what course an encounter between two cichlids belonging to different species will take. If their appearance is radically different, they will—assuming they are of the same size—often ignore each other. However, members of the same species, or fish that look as though they belonged to the same species, often will engage in aggressive fighting behavior. Especially if the two fish are males, rivalry fights may erupt. In many cases, coloration or color patterns can change within seconds, which represents a considerably more sophisticated stage of color signaling than what we usually think of when we hear the term "color pattern." This kind of "information" is concerned primarily with moods, such as aggression or preference for flight, but it also can convey readiness to mate and spawn.

A dramatic example of the effect that coloration used for signaling can have is given by cichlid species in which only the strongest male present in a territory (this can be observed in aquariums, too) exhibits the typical male colors in full brightness. *Aulonocara jacobfreibergi* is a case in point. Here all the other sexually mature males present in the same territory take on female coloring. However, if the resplendent dominant male is removed, the next ranking male quickly will develop the bright male color pattern and take on the role of dominance connected with this coloration.

Typical changes in coloration take place primarily in emergency situations. When in extreme discomfort, for instance, cichlids take on their so-called fright colors. Cichlids also change color dramatically during brood care. A change of colors can be observed most easily at night. If, one night, you suddenly turn on the tank lights, you will catch your cichlids in their "nighttime attire," which in most species is very close to their fright coloration.

Sound Utterances and Chemical Signals

Cichlids also are capable of producing sounds, as has been known for some time. These sounds usually are uttered in connection with aggressive behavior. But the exact meaning of these sounds has not yet been studied extensively and is a subject that no doubt will be explored with great interest by ethologists. Even less is known about the meaning of chemical signals, although it has been demonstrated that female mouth brooders use chemical signals to indicate readiness to spawn.

Behavior Patterns

Cichlids are known for their complex behavior patterns, which are complemented by the methods of communication already discussed. These behavior patterns are associated with the different spheres of activity that make up the life of a cichlid. Feeding behavior already has been described in the chapter, Proper Nutrition (see page 25). Just as important are behavior complexes that have to do with the interaction of cichlids among themselves, that is, behavior connected with shoaling, establishing territory, and reproduction.

Shoaling Behavior

Many cichlids are territorial only during their reproductive period and are otherwise definitely shoal or group oriented in their behavior. Shoaling behavior can be observed especially well in young fish. The shoal is of considerable protective significance for young fry. A predator contemplating attack finds it hard to concentrate on one particular fish in a shoal. Thus each individual fish has a better chance of escaping the enemy. Fully grown cichlids, too, seek the proximity of others of their kind when they are frightened or suspect danger.

Territorial Behavior

Cichlids give up their shoaling behavior at the latest when they establish their territories. Now they begin to adopt aggressive behavior to keep competitors out. Territories can be divided into three types according to the function they serve. There are feeding territories, brooding territories, and mating territories. Many cichlids living in the wild use the same territory for all three activities.

Feeding territories vary in size depending on the particular species and on its feeding habits. Fish that have specialized in feeding on algae, such as the *Pseudotropheus* cichlids of Lake Malawi, defend their feeding territories bitterly against others of their own kind who, after all, are direct competitors for the same food. Algae are rather low in food value. If the fish did not defend their feeding territory vigorously, they would starve as a consequence of having to share the scanty food supply with too many others. Cichlids that do not establish feeding territories usually rely on food sources that cannot be defended. One such group of cichlids — not discussed in this book — is the carp cichlids (some *Tilapia* species) from Lake Tanganyika. These fish live on plankton that floats suspended in the water.

Brooding territories are established and aggressively defended by almost all cichlids. The same is true of mating and spawning territories, which are discussed in the chapter, Breeding Cichlids (see page 40).

Aggressive Behavior

Fights between unevenly matched opponents usually are over quickly. A fish that clearly is stronger will quickly drive off an adversary without having to resort to biting. The danger of prolonged, violent fights is more likely in a conflict between more evenly matched fish. These encounters generally start with mock or symbolic fighting, the so-called display behavior.

Cichlids fighting mouth to mouth. This is the most aggressive phase of fighting. The cichlids grab each other by the lips and engage in a violent "tug-of-war."

Display behavior: The adversaries fan out their fins and lower the bottom of the mouth in an exaggerated fashion, making themselves "look big," and also take on their brightest colors. At the same time, one hovers broadside to the other. Now the tail is beaten to send a wave in the direction of the opponent. This is called a lateral display (see drawing in right-hand column). Another threatening stance is the frontal display. Here the gill covers are spread wide, too, and the bottom of the mouth lowered to increase the appearance of the head. Quite often, fighting cichlids now try to ram into each other. This marks the transition from symbolic to aggressive fighting. Things get to this point only if the previous, essentially symbolic fighting behavior has not brought about a decision, with the loser either making his escape or accepting a subordinate position.

Mouth fights: In this form of contest (see drawing), adversaries grab each other by the mouth and pull and shove each other around in the water. As in the case of ramming, fish can do considerable harm to each other in such a mouth-to-mouth contest. If there was any question left before, this phase of the battle definitely will settle the final outcome.

The details of fighting behavior vary, of course, from species to species, but the basic pattern is the same.

Caution! In the narrow confines of an aquarium, the defeated fish cannot get away as far from his enemy as he would in nature. If he cannot find a place to hide, his mere continued presence is regarded by the victor as a challenge to go on fighting. Thus, whereas in nature the flight of the subordinate fish means that there will be no further encounters, similar contests in an aquarium can lead to a deadly end. That is why the aquarist has to observe conflicts in the tank especially closely and assist if necessary, perhaps removing the subordinate fish, or constructing additional hiding places for the loser.

Display behavior in cichlids: When trying to intimidate each other, cichlids often assume a typical broadside position. This so-called lateral threat posture is still part of the harmless, "symbolic" stage of fighting.

Breeding Cichlids

Watching the reproductive behavior of cichlids is one of the most fascinating experiences that the hobby of keeping tropical fish offers. At first, baby fish almost always are the result of chance breeding, but after a while many aquarists aim at breeding specific kinds of cichlids. Before going on to give practical advice on how to breed cichlids, I should like to provide you with some information on the various habits of spawning, mating, and parental care cichlids have evolved, an area of behavior in which cichlids have developed more variety than any other kind of fish.

Substrate and Mouthbrooders

Cichlids are divided into two groups based on their spawning behavior.

Substrate brooders attach their eggs to a surface. The so-called open-water brooders among them use either large leaves or branches submerged vertically in the water to spawn on, or they lay their eggs on rocks. The well-known angelfish (*Pterophyllum scalare*) belongs to this category. There are open-water brooders, like the African *Anomalochromis thomasi*, that pick out individual stones, or sometimes pieces of wood, lying on the bottom to spawn on.

Other substrate brooders, called secret, hidden, or concealed brooders, deposit their eggs in larger or smaller caves formed by piles of stones or hollow roots. There are also many cichlids that prefer to dig holes themselves for spawning. They usually do this underneath flat rocks that lie directly on the bottom. Most of the cichlids that are secret brooders are rather small, and many are referred to as dwarf cichlids. A great number of secret brooders belong to the genera *Apistogramma* (South America), *Pelvicachromis* (West Africa), and *Neolamprologus* (Lake Tanganyika).

Mouthbrooders take their eggs into their mouths, usually right after laying them (ovophile mouthbrooders). Most mouthbrooding cichlids follow this behavior, among them *Pseudotropheus zebra* from Lake Malawi. But in West Africa and South America there are also a few species of mouthbrooders that first deposit their eggs on a substrate and don't pick them up in their mouths until they have hatched into larvae (larvaphile mouthbrooders). These fish thus represent an intermediate stage between substrate brooders and mouthbrooders.

Courtship Behavior

Before cichlids form a shorter or longer lasting pair bond, they usually go through an extensive courtship. The purpose of the courtship display is to advertise that the courting fish is a possible partner for reproduction or, if a bond already has been formed, to synchronize sexual readiness before actual spawning takes place. With substrate brooders, certain greeting ceremonies, trembling of the body, or, sometimes, the cleaning of possible spawning sites form part of the courtship ritual. Depending on the species, one sex courts more actively than the other. Thus, in many secret brooders the female takes the active part of courtship. Female *Pelvicachromis* cichlids from West Africa, for instance, display the beautiful red color on their bellies by presenting their bodies, bent in a U shape, to the male. In species without significant sexual bimorphism, the two sexes usually take about equal part in the courtship.

The courtship of mouthbrooders differs from that of sustrate brooders in that it is the male that intensively courts the inconspicuous looking female. He establishes special courting and spawning territories that serve this purpose exclusively. The courtship ritual usually consists of nothing more than the brilliantly colored male shaking himself in front of the drab female and leading her toward the spawning site he has chosen.

Spawning Behavior

In both substrate and mouthbrooders, the actual spawning is preceded by so-called symbolic spawning. Symbolic spawning follows exactly the same pattern as actual spawning, the only difference being that in the former no eggs and sperm are produced.

In substrate brooders the female always deposits a certain number of eggs on the substrate, to which they adhere. Then the male fertilizes the eggs by releasing his sperm into the water over the substrate. The females of secret brooders are often much smaller than the males. So when a female swims into a brooding cave, the male cannot follow because he is too big. That is why the males usually give off their sperm in front of the cavity. They then set the water moving with their fins to wash the sperm cells into the cavity, where they fertilize the eggs. Batches of eggs are laid and then fertilized in this manner until the female has exhausted her egg supply.

With ovophile mouthbrooders (see page 40), spawning happens in the following manner. The fish circle around each other, and the female proceeds to lay her eggs, picking them up in her mouth before they are fertilized by the male. Fertilization does not take place until the eggs are inside her mouth. In some highly specialized mouthbrooders it works like this: The female swims up to the male in so-called T-position. As she does so, the male spreads his caudal fin, which often has spots that look like eggs, and at the same time releases sperm. The female now tries to gather up what look to her like eggs on the male's caudal fin, and in the process sperm enters her mouth, where it fertilizes the eggs. After having spawned, the female leaves the courtship and spawning area of the male and broods the eggs by herself.

The Eggs of Cichlids

The size and number of eggs depends on several things. One factor is that the size of the eggs increases and their number decreases in proportion to the degree of protection afforded by specialization and intensity of parental care that a particular species offers its offspring. Thus, highly specialized mouthbrooders like the *Tropheus* cichlids from Lake Tanganyika lay only 10 to 20 eggs that are about 1/4 inch (6–7 mm) long. Other cichlids, especially open-water brooders, may lay hundreds or even thousands of eggs at one spawning. These eggs are tiny, and the fry that hatch from them are much less fully developed and less independent than fry hatched from larger eggs. Open-water brooders compensate for the greater mortality rate inherent in the more dangerous spawning environment by producing larger numbers of eggs. In the case of secret brooders, which spawn in caves and therefore cannot afford to increase their body size, the number of eggs can be increased only by producing smaller ones.

The color of the eggs also varies with the species. Yolk-colored and green eggs are common in species that brood in seclusion, that is in mouthbrooders and substrate spawners that are secret brooders. Color doesn't have to serve for camouflage. The eggs of open-water brooders, on the other hand, usually are beige or mud colored and thus hardly visible against their background.

Note: The kind and amounts of food eaten by the parents can produce slight variations in the color of the eggs.

Parental Care

Almost all cichlids engage in intensive parental care. First, the eggs are guarded or taken care of until they hatch.

After the hatching of the larvae (many cichlids aid the hatching process by biting off the egg shell) the parent fish go on looking diligently after their offspring. In some cases, the still immobile larvae are moved repeatedly to new shelters. There are probably hygienic reasons for his behavior. New hiding places are clean, and the larvae are thus less likely to be exposed to harmful bacteria or fungi that might have developed in the old shelter. Once the larvae have used up the food in their yolk sac, which sustains them for the first few days, they start swimming freely, that is, they now are able to move around in open water and look for food there.

Once the fry are able to swim freely, the parents take them along when foraging for food, though sometimes the parents follow behind the fry traveling in search of food. A parent will pick up individual fish that swim away from the shoal in the mouth and bring them back to the shoal, where they are spit out again. Sometimes the parents actively help provide food for their offspring. They chew large chunks of food into small bits and spit them into the shoal of fry, which gobble up the food eagerly. This period of intensive brood care lasts for several weeks and sometimes even months with open-water substrate spawners. With secret brooders and mouthbrooders, this phase is considerably shorter. The young fish become independent earlier and move away from their parents.

Telling the sexes apart. In the genus *Lamprologus callipterus* from Lake Tanganyika, the female is much smaller than the male.

Division of Labor in Brood Care

Which sex takes on the parental tasks varies from species to species. Ethologists speak of different family patterns in this connection.

Parental family: In this family pattern, which is probably the first one to evolve among cichlids, both parents share equally in the care of the offspring. Flag cichlids (*Laetacara curviceps*) follow this pattern.

Father/mother family: Here, too, both sexes take part in parental care, but they perform different tasks. Often the male defends the territory while the female tends to the offspring. Kribensis (*Pelvicachromis pulcher*), as well as other *Pelvicachromis* species, are typical examples of the father/mother family pattern.

Male-with-harem family: This pattern exists where a male gathers several females in one territory and then spawns with them one after the other. The distribution of parental chores is organized as in father/mother families. The male-with-harem family pattern can be observed in only a few aquarium cichlids, such as some South American *Apistogramma* species and some West and East African substrate spawners.

Mother family: Here the mother takes sole responsibility for the well-being of the offspring. Most mouthbrooders belong to this family type, for the female leaves the male's courtship and spawning territory after the eggs are laid. As far as we know, all cichlids from Lake Malawi adhere to this family form, as do ''*Geophagus*'' *steindacheri* from South America. Other family patterns do occur among mouthbrooding cichlids but are rare.

Extended family: In extended families, not only the parents are involved in brood care but older siblings from previous spawnings as well. Thus it can happen in an aquarium that the parent fish will allow older offspring to stay in the territory as long as they do their share in caring for their younger siblings. This family pattern is found among some secret brooders from Lake Tanganyika belonging to the genera *Julidochromis* and *Neolamprologus*.

Practical Advice for Breeding Cichlids

I cannot give you any simple, practical advice for breeding cichlids along the lines of "First do this, then that . . . ." On the contrary, quite different paths can lead to success.

Unplanned breeding: Cichlids are by nature so eager to reproduce that often they cannot be prevented from breeding in an aquarium. Once two cichlids have paired up, they will spawn in a community or a species tank without the aquarist doing anything special to encourage them. But in a community aquarium, the presence of a pair of cichlids intent on mating can be anything but pleasant for the other tank inhabitants because cichlids are exceptionally aggressive toward other fish during courtship, spawning, and rearing. And because of the large number of offspring, unplanned reproduction often is more of a problem for the aquarist than a cause for rejoicing. On the other hand, the problem usually takes care of itself because other fish in the tank may eat the fry or the parents may neglect their brood so that the baby fish die.

Planned breeding: Hobbyists interested in nature should plan to breed their cichlids in order to have the opportunity of observing the fascinating reproductive behavior of these fish. Another reason for breeding one's fish is to help make some rarer or exceptionally interesting species become better known and more readily available. Cichlids that are not stocked routinely by dealers especially, can be passed on to other hobbyists, with the possible pleasant side benefit that the transaction may recompense the hobbyist for some of the cost involved in keeping fish. This is the exception rather than the rule, however, and nobody should count on a financial return.

Where to Get Fish for Breeding

First method: Buy 6 to 12 young fish (2/3 females and 1/3 males) and raise them under optimal conditions (see page 48). When the fish reach sexual maturity, they will form pairs, and once a pair begins to engage in courtship behavior, you should move it to a breeding tank (see below).

Second method: In the case of expensive cichlids, it occasionally may be advisable to buy fish for breeding in pairs. However, there is no way of knowing for sure whether you have acquired a "real pair" until after the purchase. For hobbyists with little or no experience with cichlids, sexing these fish is not all that easy, and even if you succeed in getting a male and a female, this doesn't necessarily mean that you have a congenial pair. The greatest danger inherent in this method is that fish combined this way will not be compatible and will refuse to make friends (see Fights Between Cichlid Partners, page 48).

My tip: I generally use the first method. Resort to the second method only in special situations as, for instance, if you happen to come upon a pair of cichlids making spawning preparations in a dealer's tank.

The Breeding Tank

It is always a good idea to set up a special breeding aquarium for the cichlids you would like to breed. Place the breeding pair or, with species that do not form pair bonds, a breeding group consisting of one male and three females into this tank.

Important: Watch the fish carefully for several days after relocating them. During this critical phase, the fish may engage in violent fighting. This can happen even with pairs that were getting along fine before (see Fights Between Cichlid Partners, page 48).

The technical equipment for a breeding tank essentially should be the same as for a normal fish tank. One thing to watch out for is that the openings of the filter entrance are small enough that young fry cannot be sucked into the filter. A layer of foam rubber or a plastic shield with tiny holes makes a good barrier. Another thing that is essential is that the fish have the proper bottom material (see the descriptions starting on page 49). For many substrate spawners, for instance, it is important that the bottom material not be too coarse. Otherwise the tiny fish larvae may fall between the pieces of gravel, where the parents cannot reach them.

Arrange plenty of hiding places for the female where she can get away from the male, who is usually larger. Make sure, also, that there are several spawning sites (see page 47). Don't forget to arrange for territorial borders and visual screens. They can consist of rock structures, pieces of wood, single plants, or plant clusters. When you plan all this, remember that the tank should be easy to clean. And, finally, don't forget to place a glass cover on the tank.

The water should be especially clean, and an efficient filter therefore is important in a breeding tank. Different species of cichlids require different water properties. Exact requirements for the degree of acidity or alkalinity (pH value) and for hardness (dH) are given in the descriptions starting on page 49.

As location you should choose as quiet a place in your home as possible. Too much disturbance, such as people walking by constantly, can be upsetting enough to cause fights between parent fish tending their brood.

Spawning Sites for Substrate Brooders

Flat rocks are preferred by many open-water brooders, such as *Anomalochromis thomasi*. Place the rocks in such a way that they are distributed on the bottom and protected by nearby rock structures, individual plants, or plant clusters.

Pieces of slate are also very popular with open-water brooders. The fish spawn against pieces of slate placed vertically in the tank.

Pieces of wood (collected from bogs) also can supply good spawning surfaces for open-water brooders.

Caves for hidden brooders. Some cichlids deposit their eggs in secluded cavities. In an aquarium, a clay flower pot or a coconut shell split in half with an entrance hole chipped out can serve as a spawning and brooding site.

Caves formed by stones and accessible only from one side through a small opening are favorite spawning sites for secret brooders.

Halves of coconut shells with an entrance hole chipped out (see drawing) also can serve as protective brooding caves.

Clay flower pots are placed on the bottom upside down. Equipped with an entrance hole (see drawing), they function very well for spawning. Many species, such as *"Cichlasoma" nigrofasciatum*, also will spawn in flower pots that are split in half.

Cichlids from Central America. ▶
Above left: Salvin's Cichlid *("Cichlasoma" salvini)*; above right: Spilstrum *("Cichlasoma" nicaraguense)*; middle: Firemouth *(Thorichthys meeki)*; below left: Red-headed Cichlid *(Paratheraps* (formerly *Cichlasoma) synspilum)*; below right: Convict Cichlid *("Cichlasoma" nigrofasciatum)*.

Conical clay vases (see drawing) are useful particularly for raising discus fish. Their use is recommended primarily for hygienic reasons. These vases are available from florists who sell them for use in cemeteries. In nature, discus fish spawn on submerged tree trunks and limbs.

Large-leafed plants (*Echinodorus, Cryptocoryne*) are preferred for spawning by some cichlids, such as angelfish.

Note: A pair of cichlids occasionally will spawn directly against the glass walls of a tank. This behavior no doubt is caused by a shortage of adequate spawning sites. I therefore recommend that you set up several spawning sites to give the fish plenty of choice.

Spawning Sites for Mouthbrooders

Mouthbrooders generally are not very particular about spawning sites, but many clearly like flat sandy areas best where they can dig small pits. However, most species also will spawn on a gravelly bottom and on flagstone. It is a good idea to locate such spawning sites in the protective vicinity of rock structures or plants.

How You Can Encourage Spawning

In nature there are certain conditions and processes that have a beneficial effect on the fish's ability to spawn or that actually trigger spawning. A heavy rain can have this effect, for instance, by raising the water level and altering water quality. Most cichlids living in moving water spawn at the beginning of the rainy season. You can simulate this natural process in an aquarium by doing the following:

- Change a third of the water in the breeding tank several times a week.
- Recreate as closely as possible the natural environmental conditions of the particular cichlids in your tank.
- Offer the parent fish varied live foods of the kind most suitable for them, and feed them several times a day.

Note: Overfeeding fish makes them fat and often has a negative effect on their readiness to spawn. Feeding fish several times a day therefore doesn't mean stuffing them so that their bellies are always near bursting.

- Add two hardy fish (see companion fish, page 10) to the tank with the pair of cichlids. The latter, regarding these fish as intruders, will stick togeth-

Spawning cones: The clay vases sold by florists for use in cemeteries make a very good spawning substrate for breeding discus fish.

◀ Pair-forming hidden brooders. "Red" Kribensis *(Pelvicachromis pulcher)* from Nigeria with fry. The male defends the brooding territory, as shown here, while the female looks after the eggs and larvae. When the fry start swimming freely, both parents accompany the young.

er even more closely and later look after the brood very diligently.

With persistently uncooperative fish, that is, if the pair gets along without conflict but refuses to spawn, even minor changes can help, such as:
• the raising of water temperature slightly;
• a combination of slightly cooler water, reduced food, and less frequent water change, followed by a sudden change to optimal conditions; and
• the company of another pair of cichlids that spawns reliably and regularly (only recommended for a large breeding tank).

Fights Between Cichlid Partners

If a cichlid pair starts fighting, the fish usually have to be separated from each other. You can do this by dividing the breeding tank in two with a pane of glass. This way the fish can see but not harm each other. Put the aggressive partner into the smaller half that preferably should have no hiding place. This leaves the weaker party in a position of superiority. Watch the fish for several days, then remove the dividing glass. The pair may get along better now. If there is renewed fighting, separate them again, and repeat the process until they coexist peacefully. With a pair that got along well earlier, a renewal of the bond is likely. But there are pairs that will not accept each other again. In that case, you have to find new partners. Conflict between a pair of cichlids also can erupt in later phases of the reproductive cycle. The procedures to try to restore harmony are the same.

Raising Young Fry

As soon as the fish larvae have consumed the contents of their yolk sacs and swim freely, they need to be given food. The kinds of food suitable for them are described in the chapter, Proper Nutrition (see page 25).

There are a few rules that should be observed:

• Feed several times a day (at least three or four times); young fry store no calories because all their energy goes into growth.
• Especially with large shoals of fry, it is important, from about one to two weeks after the fish start to swim freely, to change the water frequently. The consumption of large amounts of food produces more organic waste products, which, combined with uneaten food, quickly contaminate the water. Dirty water inhibits the growth of the young fish and can result in dwarfed and misshapen fish.
• If the fry have grown at an uneven rate and some are larger than others at a certain point, the different sizes should be separated from each other. Otherwise the larger fish dominate over the smaller ones and further inhibit their growth (stress).

My tip: It is better to raise fewer fish from a large brood than to try to raise too many in too small a space.

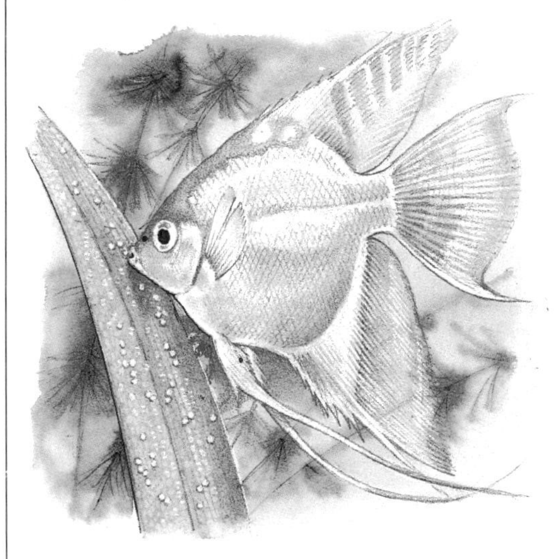

Angelfish engaged in brood care. The angelfish is a typical open-water brooder that likes to deposit its eggs on large plant leaves. As part of their conscientious brood care, the parents remove dead and fungus-infected eggs.

Popular Cichlid Species

The earlier chapters of this book were devoted to general information on how to keep cichlids in an aquarium. The following descriptions of popular cichlids suitable for beginning hobbyists contain more detailed data on the appearance, environmental conditions, care, and breeding of some species. All the information given is based on my own observations and experience. Deviations are, of course, possible in some cases, but they are on the whole insignificant, as I have found in the course of many discussions with experienced aquarists.

Some Notes on the Descriptions

The Latin names of some cichlid species recently have been changed. In addition to the names reflecting the latest state of scientific practice I also include older names, but only if a species has been traded under its older name for some time. Let me give an example: *Laetacara curviceps* (the flag cichlid) also is known as *Aequidens curviceps*; therefore the species is listed under both Latin names.

The suggestions for the tank interior refer to the table Model Communities (see page 12).

Species with similar needs are listed at the end of some of the descriptions. If these species differ in some respects from the cichlids discussed in that section, this fact also is mentioned.

Cichlids from West, Central, and East Africa

(not including Lakes Malawi and Tanganyika)

Cichlids from these parts of Africa are very popular with hobbyists. One reason is clearly that many of these fish, particularly the West African ones, are small or at most of medium size. Another reason is that many very colorful dwarf cichlids whose care is not especially demanding come from those regions. Only cichlids belonging to the genera *Tilapia, Sarotherodon*, and *Oreochromis* are regarded by many hobbyists as problematic aquarium fish. In my opinion this negative assessment is unjustified because these are robust fish, although they do grow rather large, reaching a length of over 8 inches (20 cm).

Anomalochromis thomasi

Formerly: *Pelmatochromis thomasi*

Geographic origin: Sierra Leone, Liberia.

Size: 2¼ to 3 inches (6–8 cm).

Appearance: Colorful small cichlid, rather high-backed, with many rows of shiny scales on the sides and two dark spots on the flank and caudal peduncle; different color variants. Most fish sold are tank-bred.

Sexual differences: Hard to detect even in adult fish; females sturdier, rounder, and with more dark markings.

Behavior: Forms pair bonds and establishes territories; peaceful and therefore easy to combine with other fishes.

Care: One or two pairs in a community aquarium containing African cichlids (aquarium types 1 and 2, page 12).

Water: 75° to 81°F (24–27°C); pH 7; 7–15°dH.

Feeding: All types of food described, in medium size morsels.

Breeding: Open-water spawner; offer it several flat stones for spawning. The eggs often are eaten; occasionally two females of this species spawn together.

My tip: Even if breeding efforts are unsuccessful, this species is recommended highly for community tanks.

Chromidotilapia guentheri

Guenther's Mouthbrooder

Photo on page 17

Geographic origin: From Libya to Cameroon.

Size: Females up to 6 inches (15 cm), males up to 7 inches (18 cm); may be smaller, depending on what bodies of water they come from originally.

Appearance: Pointed mouth, laterally compressed body. Reminiscent of South American *Geophagus* cichlids.

Sexual differences: Females more colorful with a shiny longitudinal band on the anterior part of the dorsal fin and a red abdomen.

Behavior: Quite peaceful until sexual maturity and pair formation, at which point these fish can start to act up and often chew or dig through the bottom material; relatively shy.

Care: Can be kept in a community tank (aquarium type 2, page 12) with other medium-sized West African cichlids until spawning time. Fish that are to be bred preferably should be in a tank of their own.

Water: 75° to 82°F (24–28°C); pH 6.5–7.5; up to 18°dH.

Feeding: All types of food described; prefers live and frozen food.

Breeding: Mouthbrooder but pair forming. Eggs are picked up in the mouth by the male directly after being laid; later the female helps care for the hatched brood and also takes fry into her mouth. Should be raised on Artemia nauplii.

Species with similar requirements: All other *Chromidotilapia* species. *Thysochromis ansorgii*, a substrate spawner (see page 44), is more placid.

Hemichromis cristatus "Red Cichlid"

Crown Jewel Cichlid

Photo on inside front cover.

For a long time red cichlid was a term commonly used for all *Hemichromis* species except *H. fasciatus* and *H. elongatus*. Most of the cichlids of this genus are sold by dealers as *Hemichromis lifalili* or *H. bimaculatus*.

Geographic origin: West and Central Africa.

Size: 3 to 4 inches (8–10 cm).

Appearance: Orange to red body with one dark mark each on the gill cover and on the middle of the body; rows of bluish iridescent dots on the body and the fins. Depending on the species, the red is more or less bright and the marks and iridescent dots more or less prominent.

Open-water brooders. This kind of cichlid lays its eggs not in secluded caves but in the open, on flat stones, pieces of wood, or leaves of largish plants.

Sexual differences: Difficult to see. Males are slightly larger, females, somewhat rounder.

Behavior: Robust, aggressive cichlid; hard to combine with other fish, especially during spawning time; likes to burrow.

Care: Needs a large tank (at least 40 inches [100 cm] long for a pair); not demanding as far as the interior of the tank is concerned.

Water: 75° to 82°F (24–28°C); pH 6–7.5; 4–20°dH.

Feeding: All types of food described; prefers live and frozen food.

Breeding: Simple; open-water brooder that spawns on stones or pieces of wood; lays several hundred eggs and attends to the fry for a long time.

My tip: A delightful fish for cichlid fanciers who enjoy devoting time to their fish. Highly recommended in spite of the numerous drawbacks.

Watch very carefully to make sure a pair doesn't begin to fight (see page 48).

Species with similar requirements: All *Hemichromis* cichlids require very similar care; some, especially *H. fasciatus*, become somewhat larger.

Pelvicachromis pulcher

"Kribensis"

Photo on page 16.

Formerly: *Pelmatochromis pulcher, Pelmatochromis kribensis*

Geographic origin: Nigeria.

Size: 3 to 4 inches (8–10 cm).

Appearance and sexual differences: Longish oval shape; males have a somewhat more elongated body and a caudal fin that is long and pointed. Females are shorter, more stocky, and usually more colorful with a red area on the abdomen. There are several different color varieties. Depending on the variety, there may be eye-shaped markings on the dorsal and/or caudal fins and a longitudinal band on the body.

Behavior: Pair-forming hidden brooders, yet can easily be combined with other fishes.

Care: One or two pairs in a community tank containing African cichlids (aquarium types 1 and 2, page 12); likes a coconut shell split in half and equipped with a small entry hole (1 inch [2.5 cm] wide) for brooding.

Water: 75° to 81°F (24–27°C); pH 6.5–7.5; 7–15° dH.

Feeding: All types of food described, in medium-sized morsels; prefers live food.

Breeding: Raise water temperature to 81° or 82°F (27–28°C); water should be soft and slightly acid (pH just below 7, dH up to 15). Female looks after eggs and larvae while male guards territory; both parents guide the young.

My tip: A gorgeous and ideal cichlid that is suited for beginners.

Species with similar requirements: All West African hidden brooders of the genera *Pelvicachromis* and *Nanochromis*. However, these fish generally are harder to please in terms of water quality and food. Fish caught in the wild also are more susceptible to disease.

Pseudocrenilabrus multicolor

Egyptian Mouthbrooder

Other names: *Hemihaplochromis multicolor, Haplochromis multicolor*

Geographic origin: North and East Africa

Size: Males up to 3 inches (8 cm), females up to 2½ inches (6 cm).

Appearance: Elongated oval body with scales that have a metallic greenish-bluish sheen; fins have a pattern of orange. The tank-bred fish one finds at pet stores often lack the beautiful colors of their wild ancestors.

Sexual differences: Males are larger and more colorful; females are smaller and paler, with dark markings. Males have red on the end of the anal fin.

Behavior: Males tend to fight with each other but get along with other species; don't bother plants.

Care: Can be kept in relatively small tanks (at least 32 inches [80 cm]) with plants along the walls (see Aquarium Plants, page 23) and lots of hiding places; sand or gravel bottom. Combine one male with two or three females.

Water: 72° to 81°F (22–27°C); pH 7; 10–15° dH.

Feeding: All types of food described but prefers live food; small morsels.

Breeding: Mouthbrooders that do not form pair bonds; up to 60 eggs. Fry are still cared for after they first are released.

My tip: Make sure the fish you buy have good coloring; in dealers' tanks, females often breed when they still are much too small.

Species with similar requirements: *P. philander dispersus*. This species usually grows to a larger size and needs correspondingly larger tanks. The same applies to *Astatotilapia* and most *Haplochromis* species from Lake Victoria. Fish sold by dealers under the name of *Astatotilapia burtoni* (formerly ''*Haplochromis*'' *burtoni*) presumably are various species of *Pseudocrenilabrus* with similar requirements. *P. nicholsi* is a very beautiful cichlid that dealers have begun to sell only recently and that may in the future largely replace *P. multicolor* as an aquarium fish.

Steatocranus casuarius

Lionhead, Blockhead, or Lumphead Cichlid
Photo on inside back cover.

Geographic origin: Rapids in the lower Zaire River (Congo).

Size: Males, up to 4¼ inches (11 cm); females, up to 3 inches (8 cm).

Appearance: Oval, laterally compressed cichlid; ground color light beige to dark brown, depending on mood.

Sexual differences: The males, which develop a large hump on the forehead as they grow older, are larger than the females; juveniles are difficult to sex.

Behavior: Used to river rapids, this species lives very close to the bottom. Because of their reduced swim bladder these fish are unable to float motionless in the water.

Care: Keep in a community tank (aquarium type 1 or 2, see page 12). Provide many hiding places in rock piles or under horizontally placed, flat rocks. Plantings possible. Combine with other species (such as *Chromidotilapia* cichlids or *Anomalochromis thomasi*) or keep a single pair in a breeding tank at least 20 inches (50 cm) long.

Water: 75° to 82°F (24–28°C); pH 6.5–7.5; up to 15°dH.

Feeding: All the various types of dry, live, and frozen food described.

Breeding: Fairly easy; pair-forming secret brooders; aggressive during spawning time. Rear young fish on Artemia nauplii. The parent fish look after their young but not very assiduously.

My tip: It is safer not to combine this species with other hidden brooders (*Pelvicachromis*) or with other cichlids from river rapids (*Teleogramma*) because they compete for the same spawning caves.

Species with similar requirements: Other *Steatocranus* species, especially *S. tinanti*. *S. ubanguiensis*, which is still relatively unknown, looks very much like *S. casuaris* but is smaller. Brichard's Teleo (*Teleogramma brichardi*) is also similar. *Lamprologus congoensis*, which tends to form harems, lacks striking coloration but is a cichlid well worth keeping.

Cichlids from Lake Malawi

Malawi is home to about 500 cichlid species, many of which have not yet been described scientifically (the same applies to the cichlids of Lake Tanganyika). The water in this lake is clear and very uniform in quality. All cichlids endemic to Lake Malawi are mouthbrooders. They can be divided into three groups: rock-dwelling cichlids that feed on algae and sometimes are referred to as ''Mbunas,'' the genera *Aulonocara* and *Trematocranus*, and the large group of *Haplochromis*-type cichlids, including the ''Utaka,'' which have specialized in eating plankton. Many cichlids belonging to the first two groups tend to develop different colors and other local variations.

Aulonocara nyassae

Nyasa Peacock
Photo on page 28.

Geographic origin: Lake Malawi.

Size: Up to 7 inches (18 cm); often smaller in aquariums.

Appearance: High-backed, laterally very compressed.

Sexual differences: Males more brightly colored; females relatively pale and somewhat smaller.

Behavior: Quiet and peaceful; although these fish form territories, they are not aggressive even toward their own kind.

Care: A group of one male and three to four females in an aquarium of type 6 (see page 13); 2

males and up to ten females can be kept in a tank of at least 50 inches (130 cm) and with lots of retreats; plantings are possible.

Water: 77° to 82°F (25–28°C); pH 7.5–8; 10–15°dH.

Feeding: All types of food described; prefers live or frozen food.

Breeding: Specialized mouthbrooders, not pair-forming. Like to spawn in caves on a stone bottom; 40 to 50 eggs. The young fish are released after about three weeks.

My tip: Make sure you always buy the females together with the males to make sure you have fish of the same species and to prevent hybridization. Females of the various *Aulonocara* species look very similar.

Species with similar requirements: All *Aulonocara* cichlids (*A. baenschi* and *A. maylandi*, for instance) need essentially the same care. *A. (''Trematocranus'') jacobfreibergi* is somewhat smaller but particularly beautiful.

Copadichromis boadzulu

Other names: *Cyrtocara boadzulu*, *Haplochromis boadzulu*

Geographic origin: Lake Malawi.

Size: 4¾ to 6 inches (12–15 cm); females somewhat smaller.

Appearance: High-backed and laterally very compressed. Males basically are blue with scales in the lower and posterior area of the sides that are rimmed with orange to red and with longitudinal stripes of the same color on the bluish tail fin.

Sexual differences: Females somewhat smaller, inconspicuous silvery gray, with two dark longitudinal stripes.

Behavior: Quiet and peaceful, even shy if together with restless fish; in tanks with several males, only one displays full coloring.

Care: Keep in an aquarium of type 6 (see page 13) that is at least 48 inches (120 cm) long and that should be well planted but still have enough open area for swimming. Anchor plants well in bottom. Caves and rock structures are not as important to the well-being of these fish.

Water: 73° to 81°F (23–27°C); pH above 7.5; 12° dH or higher.

Feeding: All types of food described; prefers small live food animals, such as water fleas. In their natural habitat, these fish eat plankton.

Breeding: Non-pair forming mouthbrooders typical of *Haplochromis*-type cichlids from Lake Malawi, but quieter. Keep one male together with three or four females. Breeding efforts are foiled easily if the need for peace and quiet of these fish is disregarded.

My tip: Make sure these cichlids have enough open water for swimming, and don't combine them with other species that are too rambunctious.

Species with similar requirements: Cichlids sold in pet stores as ''*H. steveni*'' are either a color variant of *C. boadzulu* or some other closely related species. The popular ''Utaka'' cichlids have now been assigend to the genus *Copadichromis*, as, for instance, *C. borleyi*. *Cyrtocara moori* is another peaceful cichlid of this type that has similar needs, though it is not an ''Utaka.''

Many cichlids that belong to what formerly was called the *''Haplochromis''* genus also have similar requirements but generally are not quite as peaceful (for instance, *Cheilochromis euchilus, Sciaenochromis ahli, Copadichromis chrysosonotus*). Some other *''Haplochromis''* cichlids, such as *Nimbochromis livingstonii, N. polystigma*, and *Dimidiochromis compressiceps*, are downright fish predators.

Melanochromis johannii

Johannii

Photo on page 28

Geographic origin: Lake Malawi.

Size: Up to 4¾ inches (12 cm).

Appearance: Torpedo-shaped body with longitudinal bands that are not quite as well defined in the female.

Sexual differences: Males larger, with blue-

and-black longitudinal stripes; females orange, sometimes with dark longitudinal stripes.

Behavior: A typical Mbuna cichlid; territorial; does not get along with members of its own species or with cichlids other than Mbunas; active swimmer.

Care: Community tank (aquarium type 5, see page 13); one male with at least four females, together with other Mbuna cichlids.

Water: 75° to 81°F (24–27°C); pH 7; 10–20°dH.

Feeding: All kinds of live and frozen food described; give only small amounts of low-fiber food; vegetarian food is important.

Breeding: Highly specialized mouthbrooders that do not form pairs. Eggs are deposited in a shallow pit and picked up by the female in her mouth; once released, the fry are cared for only briefly.

Pseudotropheus zebra

Cobalt Blue Cichlid or Zebra

Geographic origin: Lake Malawi.

Size: 4¾ to 6 inches (12–15 cm).

Appearance: Elongated body but with a bulky head; normal variety has a dark head and numerous transverse dark stripes against light blue; many color varieties with contrasting colors especially on the unpaired fins. In addition to the different color varieties, dealers also sell cichlids under the name of *P. zebra* that really constitute separate species but have not yet been described scientifically (*P. spec. aff. zebra*, red zebra and "bright blue").

Sexual differences: Males have conspicuous and clearly defined egg spots, which the females usually lack completely. Males develop a hump on the forehead with age. In the "red zebra" variety the males usually are blue and the females orange red; in the "bright blue" variety the females are considerably paler than the brilliantly blue males. In these varieties pied females and, less frequently, pied males sometimes are encountered.

Behavior: Typical Mbuna cichlids; territorial and incompatible with others of their species. Active swimmers.

Care: In a largish tank for Malawi cichlids (aquarium type 5, page 13) keep one male and at least four females together with a number of other Mbuna cichlids. A high population density reflects the natural living conditions of Mbunas. It helps dissipate aggression, but good filtration and frequent water changes are mandatory.

Water: 75° to 81°F (24–27°C); pH 7 or higher; 10–20°dH.

Feeding: Live and frozen food; also vegetarian food in the form of food flakes with a high proportion of plant matter, or fresh spinach, lettuce, and algae.

Breeding: Non-pair forming mouthbrooders. About ten days after the eggs are laid, the fry are released and then looked after only briefly and not very intensively.

My tip: When combining Mbuna cichlids, make sure the species brought together are as different in appearance as possible (see combining fish, page 10).

Species with similar requirements: Basically all Mbuna cichlids require the same care. *Labidochromis* and many *Melanochromis* species are smaller and therefore can manage in a smaller tank; other *Melanochromis* species and *Labeotropheus* cichlids have needs almost identical to those of the species described.

Brown Discus Cichlid *(Symphysodon aequifasciatus axelrodi)* with fry. Once the brood start swimming freely, they first feed exclusively on a special skin secretion on their parents' sides. ▶

Cichlids from Lake Tanganyika

The cichlids from Lake Tanganyika fall into two major groups: Mouthbrooders and hidden brooders. Only a few kinds—which are of minimal interest to aquarists—spawn openly in sand craters. Many species of Lake Tanganyika cichlids tend to develop different color and local variations. Like Lake Malawi, Lake Tanganyika is home to several hundred cichlid species. Its water is clear and has highly uniform properties. That is why many cichlids from this lake react badly to aquarium water that is too soft or not acid enough.

Cyphotilapia frontosa

Frontosa

Photo on front cover.

Geographic origin: Lake Tanganyika.

Size: Males up to 14 inches (35 cm); females up to 10 inches (25 cm).

Appearance: High-backed, with a body depth of up to 4 inches (10 cm); with a conspicuous bulge on the forehead. Variants: 5-stripe and 6-stripe Frontosa; "Masked" Frontosa.

Sexual differences: Practically impossible to detect in juveniles; adult males have a much bigger bulge on the forehead than females.

Behavior: Quiet fish; mouthbrooder wih the female doing the brooding; non-pair forming.

Care: In spite of its size, this cichlid can be kept in tanks of 50 inches (130 cm). Provide a sandy bottom and rock structures that form caves. Plants usually are uprooted. Strong filtration recommended.

◀ Cockatoo Dwarf Cichlid *(Apistogramma cacatuoides)*. The first few rays of the dorsal fin are considerably more elongated in the male.

Water: 75° to 81°F (24–27°C); pH 7–8; 10–20°dH.

Feeding: This species eats fish, attacking sleeping fish in the semidarkness. Needs nutritious food; give it frozen food, such as fish and meat, shrimp, and beef heart; it also will happily consume food tablets in great quantities.

Breeding: These fish don't spawn until they are 1½ to 2 years old, and offspring often are not obtained until even later. Keeping several females with a male is recommended; anywhere from three to ten females can be combined with a male, depending on the size of the tank. Males often chase females but serious fighting is rare. Up to 40 eggs are laid; the fry are kept in the mouth for up to six weeks and may be .4 to .6 inch (11–15 mm) long when they are released.

My tip: The caves in the rock structures should have openings large enough for the females but too small to admit the male. Do not combine with fish that swim restlessly and eat voraciously, such as *Tropheus* cichlids. If necessary, feed individual fishes separately. Mouthbrooding females can be placed in a separate tank after about four weeks to release the young fish there. Returning them to the regular tank only rarely gives rise to problems.

Eretmodus cyanostictus

Striped Goby Cichlid

Photo on page 18.

Geographic origin: Lake Tanganyika.

Size: Up to 4 inches (10 cm).

Appearance: Longish oval body, not too compressed laterally; overshot mouth. Transverse stripes and, especially on the body, lots of light blue, iridescent dots.

Sexual differences: Almost impossible to detect in adult fish and not at all in juveniles; when fully grown, males are larger.

Behavior: Lives close to the bottom, reduced swim bladder; pair-forming mouthbrooder; restless, "hopping" way of swimming; fish often chase one another.

Care: As companion fish in community tanks (aquarium types 8a, b, and c, see page 13). Introduce several of them so that pairs can form.

Water: 75° to 81°F (23–27°C); pH above 7.5; 12°dH or higher.

Feeding: Algae eaters that react sensitively to inapproapriate diet; for details, see Feeding under *Tropheus* cichlids (page 60).

Breeding: Mouthbrooders that form pairs. Difficult to breed. Parents take turns mouthbrooding. Rearing of the fry is difficult because they stay hidden.

Species with similar requirements: *Spathodus marlieri, Sp. erythrodon, Tanganicodus irsacae.*

Julidochromis marlieri

Marlier's Julie

Photo on page 18.

Geographic origin: Lake Tanganyika.

Size: Up to 6 inches (15 cm); the tank-bred fish sold at pet stores usually are smaller.

Appearance: Elongated body, round to oval in cross section; older fish may have a slight bulge in the forehead.

Sexual differences: Not detectable in juveniles; later, the sexes sometimes can be told apart by the genital papilla, which is more pointed in the male and aimed backward; the female's pupilla is thicker and aimed frontward.

Behavior: These fish like to swim upside down with the belly facing the roof of the cave; young fish often like to fight. Later rather loose pair bonds are formed that are upset easily by external influences (changes in the tank interior, water changes); occasionally bonds occur among trios.

Care: In communal tanks (aquarium type 8, see page 13) with rock structures; for a single breeding pair a tank of 24 inches (60 cm), similarly arranged, is big enough.

Water: 75° to 81°F (24–27°C); pH 7–8; 10–20°dH.

Feeding: Live, frozen, and dry food in morsels that are not too big.

Breeding: Hidden brooders; the eggs and, later, the larvae are attached to the cave ceiling; sometimes large broods (up to 100 eggs) are produced at longer intervals and sometimes smaller broods succeed each other rapidly. The parents look after the brood, and several successive batches of offspring are tolerated by the parents and thus enjoy their protection.

My tip: Catch and remove juvenile fish now and then; otherwise the parent fish will reproduce less. Then try to restore the aquarium to its previous state as closely as possible to avoid "marital discord."

Species with similar requirements: All *Julidochromis* species, but *J. ornatus, J. transcriptus*, and *J. dickfeldi* noticeably are smaller. Also all cichlids of the genus *Chalinochromis*, although these fish are not as totally dependent on living in caves. Also *Telmatochromis bifrenatus* and *T. vittatus*, both of which take not only to cavities formed by rock structures but also to ones made of other materials (such as coconut shell halves).

Neolamprologus multifasciatus

Photo on page 18.

Geographic origin: Lake Tanganyika.

Size: Females up to 1 inch (2.5 cm); males up to 1.4 inch (3.5 cm).

Appearance: One of the smallest cichlids, with about a dozen transverse stripes.

Sexual differences: Size (see page 9); no other distinguishable differences.

Behavior: Cichlids that live in close proximity to empty snail shells, where they find refuge and also lay their eggs. They live in colonies; several males with a large number of females can coexist next to each other. They form "harem" type families and work on the snail shells that serve as their "houses" by digging up and carting away sand around them.

Care: Either keep a small colony in the foreground of a Lake Tanganyika aquarium or set up in a tank of their own measuring 20 inches (50 cm) or more. Distribute a large number of empty shells from edible snails on the fine sand of the bottom; this "snail section" may be set off from the rest of the tank with fist-sized stones.

Water: 75° to 81°F (24–27°C); pH 7–8; 10–20°dH.

Feeding: All common fish foods in small morsels. It is important to feed the growing juveniles Artemia nauplii regularly. The adult fish also happily eat the brine shrimp larvae.

Breeding: Under good care these cichlids start reproducing without special encouragement, having only a few young (two to six) at a time at first; even later they hardly hatch more than a dozen. Several batches of young fish grow up side by side within the parents' territory.

My tip: Make available as many snail shells as possible because this makes the fish dig less in the bottom. The young should not be removed until they are almost full grown, when it already is possible to recognize pairs.

Species with similar requirements: All *Neolamprologus* cichlids from Lake Tanganyika, although there are some differences to be observed. The other species grow somewhat — in some cases considerably — larger (*N. meeli, N. boulengeri* up to 3 inches or 8 cm). They do not live in colonies but almost always in pairs (*N. ocellatus, N. bervis*) and display more intraspecies aggression. In some species the male doesn't enter the snail shell (*Lamprologus callipterus*), whereas in *N. brevis* both male and female enter it. *L. callipterus* also forms harems, and there is a big difference in size between the sexes, males measuring up to 5 inches (13 cm), females growing only 1¼ or 1½ inches (3–4 cm) long (see drawing, page 42). Most *Neolamprologus* cichlids are best kept in pairs. They are close to ideal companions for all species in a Lake Tanganyika aquarium. Or a breeding pair can be kept in a tank of 20 inches (50 cm) or more. All the species display fascinating behavior patterns in the course of burying and transporting their snail shells and when spawning inside them.

Neolamprologus spec. "daffodil"

Photo on page 18.

Geographic origin: Lake Tanganyika.

Size: Up to 4 inches (10 cm).

Appearance: Elegant, with long tapered points on dorsal and tail fins. Basic color beige with yellow markings, especially on the unpaired fins.

Sexual differences: Not visible in juveniles and hard to detect in adults. Males are bigger, and the tapered ends of the fins are longer.

Behavior: Very territorial; older males and pairs with brood to care for easily become aggressive; don't bother plants.

A special group of secret brooders. In Lake Tanganyika there are many cichlids that have specialized in spawning in empty snail shells and raising their brood there. In the species *Neolamprologus boulengeri,* the female takes sole responsibility for looking after the young, whereas the considerably larger male guards the brooding site.

Care: Several fish in a community tank (aquarium type 8, see page 13) or in pairs in a breeding tank of 32 inches (80 cm) or more; use fine sand for the bottom; aquarium can be planted.

Water: 75° to 81°F (23–27°C); pH above 7.5; 12°dH or higher.

Feeding: All common types of fish food; prefers live food.

Breeding: Secret brooders that may lay batches of 150 or more eggs. Defend only a small territory around the spawning site, but the defense is vigorous. Use Artemia nauplii to feed young fry. Older batches of offspring often assist in caring for younger broods.

Species with similar requirements: All cichlids of the *N. brichardi* group, which also includes *N. spec. "daffodil"*; that is: *N. pulcher, N. savoryi, N. caudopunctatus, N. buescheri, N. spec. "walteri"*, and others. Also larger secret brooders belonging to the *Neolamprologus* genus. But these need larger caves and more nutritious food (*N. sexfasciatus, N. tretocephalus*). *Altolamprologus* cichlids are more sensitive.

Tropheus moorii

Moorii

Photo on page 18.

Geographic origin: Lake Tanganyika.

Size: Up to a maximum of 6 inches (15 cm); more commonly 4¾ to 5 inches (12–13 cm).

Appearance: High-backed, stocky body that is oval in cross section; steep profile with slightly undershot mouth.

Sexual differences: External differences hard to detect and practically nonexistent in juveniles. In grown fish the sexual papilla can give a clue (larger and rounder in females, smaller and more pointed in males), or the shape of the head and the mouth may help identify the sex; otherwise you have to wait until you can tell the gender by the courtship behavior.

Behavior: Mouthbrooders that should be kept in groups. Restless, fast swimmers that need a lot of space. Fish often chase one another and form rank hierarchies in which new fish find it hard to find a place. Not enough is known yet about behavior differences between different varieties.

Care: Keep in groups of three to four females to one male in tanks at least 48 inches (120 cm) long and as tall as possible. Build rock structures that reach close to the water surface and with many hiding places for weaker fish and brooding females. You can try to introduce hardy plants (such as Java fern).

Water: 72° to 81°F (22–27° C); pH 7–8; 12°dH or more.

Feeding: Sensitive aufwuchs feeders; like dry food, especially green flakes and tablets stuck to the glass; in addition give them frozen and live food with plenty of roughage (*Mysis*, large brine shrimp, water fleas); *avoid* red mosquito larvae, tubifex, and beef heart, because these foods quickly cause intestinal disorders in these fish.

Breeding: Non-pair forming mouthbrooders. If kept under good conditions, Moorii breed without special encouragement. Offer the females hiding places (in dense plants, piles of rocks), where they can release their five to at most 20 young. Don't take females out of the tank because the group will not accept them back after even a few days.

My tip: Don't combine with meat-eating species because the diet of the latter (for instance the large *Neolamprologus* species and *Cyphotilapia frontosa*) doesn't agree with *Tropheus* cichlids. *Julidochromis* and *Telmatochromis* species and small *Neolamprologus* are suitable companions for this fish.

Note: This cichlid comes in a large number of color and local varieties that have similar requirements; some of them have been described as separate species, such as *T. polli* and *T. brichardi*.

Species with similar requirements: *T. duboisi*, or blue-faced duboisi, which is very popular because of the unusual markings of the juveniles, is perhaps better suited to being kept in pairs.

Cichlids from Central America

Compared to cichlids from other areas of the world, the Central American cichlids form a relatively unified group. Most of them are medium large to large; in only a few species is the male less than 6 inches (15 cm) long. These fishes incorporate most completely the typical cichlid traits of a "lovable bully," combining beautiful colors and very varied behavior with rough ways and a high degree of aggressiveness especially toward others of their species.

"Cichlasoma" nigrofasciatum

Convict cichlid
Photo on page 45.

Geographic origin: On the Pacific side of Central America, Guatemala to Panama.

Size: 4¾ to 5½ inches (12–14 cm).

Appearance: Relatively high-backed, laterally compressed cichlid wearing "convict's" stripes; quite pointed, small mouth. An albino variety exists.

Sexual differences: Males are larger, with longer fin tips. Females have a shiny spot on the abdomen.

Behavior: Rather belligerent; likes to eat plants, including aquarium plants; not well suited for community tanks.

Care: One pair (aquarium type 3, see page 12) to be kept only with other, vigorous cichlids; or by itself in a species tank measuring 32 inches (80 cm) or more; let several juveniles grow up together.

Water: 73° to 81°F (23–27°C); pH ca. 7; 5–15°dH.

Feeding: All types of food described; vegetarian food is an important part of the diet. Use food flakes with a high percentage of plant matter.

Breeding: Simple; secret brooders with elements of open-water brooding; intensive brood care. Fry easy to rear with Artemia nauplii.

My tip: It is very important to try to obtain offspring from fish caught in the wild because strains that have been tank-bred over many generations tend to have poor coloring.

Mouthbrooders spawning. The male guides the female toward the spawning site with sinuous movements. Once there, the two fish circle each other, shaking violently until the female releases her eggs.

Species with similar requirements: A number of species of the *Cichlasoma* group: *C. sajica* and *C. septemfasciatum* are most ideal; also *Herotilapia multispinosa* (rainbow cichlid).

Copora nicaraguensis

Spilotum
Photo on page 45.
Formerly: *Cichlasoma nicaraguense*

Geographic origin: Nicaragua, northern Costa Rica.

Size: Females up to 8 inches (20 cm); males up to 10 inches (25 cm).

Appearance: Slender, with a steep facial profile; populations from different areas show differences in color: fish from Costa Rica have a yellow to orange body and a blue head, wheras those from Nicaragua have a beige head.

Sexual differences: Males have a slight bulge on the forehead and black reticulate markings on the anterior of the body; females have a golden dorsal fin and abdomen.

Behavior: Territorial but rather peaceful; occasionally will dig up the bottom and nibble on plants.

Care: Community tank (aquarium type 4, see page 12); you can try to introduce plants. Raise several juveniles together up to sexual maturity. Can be combined with *Cichlosoma* species of comparable size.

Water: 75° to 81°F (24–27°C); pH 7–8; 10–20°dH.

Feeding: Live and frozen food in large morsels and generous amounts; also food flakes and pellets.

Breeding: Prefers caves or hiding places for spawning, where eggs are deposited on the bottom; lengthy courtship. Several hundred nonadhesive eggs are laid, which take about three days to hatch. The fry start to swim freely about seven to eight days after spawning time. The female guides them while the male defends the territory.

"Cichlasoma" salvini

Salvin's cichlid

Photo on page 45.

Geographic origin: Southern Mexico, northern Guatemala.

Size: Males up to 6 inches (15 cm); females somewhat smaller.

Appearance: High-backed with flattened sides; yellow ground color, with two irregular longitudinal black bands running down the body and a third stripe between them on the forehead; different color varieties (reflecting locality of origin), some with vivid red on the lower flanks and on the anal and tail fins.

Sexual differences: Females smaller, with shorter, rounded fins; males with bluish to silverish shimmering scales and fins; females more brightly colored.

Behavior: Doesn't burrow in the bottom and doesn't bother plants; doesn't get along with others of its species.

Care: One pair in an aquarium of type 3 (see page 12). Planting possible.

Water: 68° to 79°F (20–26°C); pH around 7; 8–15°dH.

Feeding: All types of food described; prefers nutritious live food.

Breeding: Open-water brooders, pair forming; spawn on stones and pieces of wood. Produce several hundred eggs. Intensive parental care of the eggs and the fry.

Species with similar requirements: A number of less well-known species of the *Cichlasoma* subgenus *Parapetenia* are comparable to Salvin's cichlid in size and requirements. "*C.*" *octofasciatum* (formerly "*C.*" *biocellatum*) is the best known of them and measures about 8 inches (20 cm).

Paratheraps synspilum

Redheaded cichlid

Photo on page 45.

Formerly: *Cichlasoma synspilum*

Geographic origin: Guatemala, Belize.

Size: Up to 14 inches (35 cm).

Appearance: Bullish looking, high-backed cichlid, laterally very compressed; colorful, with reddish area on head and many black spots distributed over the body.

Sexual differences: Juveniles hard to sex; adult males develop a noticeable bulge on the forehead.

Behavior: Territorial; highly aggressive toward own species but more tolerant toward other cichlids; burrows in the bottom and likes to chew on and eat plants.

Care: Community tank (aquarium type 4, see page 12) without plants; hiding places provided by flat rocks important.

Water: 75° to 81°F (24–27°C); pH 7–7.5; 15°dH.

Feeding: Live and frozen food in large morsels and generous amounts; also food flakes and pellets.

Breeding: Trying to combine a pair for mating can be very tricky because the aggressiveness of this cichlid toward others of its kind turns meetings into dangerous events. Juveniles that have grown up together form pairs more easily. Open-water spawners that produce large numbers of young (up to over 1,000).

My tip: Full development of colors is dependent on a diet with plenty of variety.

Species with similar requirements: All cichlids belonging to the genus *Theraps* can be kept similarly. Almost all of them are brightly colored and can be recommended for beginners. *Vieja* (formerly "*Cichlasoma*") *maculicauda* is still the most commonly kept cichlid of this group.

Thorichthys meeki

Firemouth
Photo on page 45.
Other name: *Cichlasoma meeki*

Geographic origin: Mexico, northern Guatemala.

Size: Up to 6 inches (15 cm).

Appearance: High-backed cichlid that looks quite dainty, however. Throat and abdomen a beautiful red; flanks with a metallic sheen; eye spot on gill covers.

Sexual differences: Males have longer tips on dorsal and anal fins and brighter coloring.

Behavior: Territorial; during spawning season this species sometimes burrows in the bottom and then tends to be aggressive toward other fish; otherwise perfectly peaceful.

Care: One pair in a smallish tank or two to three pairs in a larger one (aquarium type 3 or 4, see page 12). Hardy plants may survive if they are protected from the burrowing fish (see page 19).

Water: 75° to 81°F (24–27°C); pH around 7; up to 12°dH.

Feeding: All types of food described; prefers live food.

Breeding: Open-water spawner; a clutch of several hundred eggs is deposited on stones; afterwards the larvae are looked after in pits; intensive brood care.

Frontal threat posture. When assuming this posture, the fish force open their gill covers and lower the bottom of the mouth to make themselves appear larger than they are. The eye spot on the gill covers, such as the one of the Firemouth shown here, magnifies the intimidating effect.

My tip: This species is highly interesting to watch; Firemouths, or one of the other very similar species (see below), are an excellent introduction to the hobby of keeping Central American cichlids.

Species with similar requirements: The cichlids now assigned to the genus *Thorichthys* used to be included in the collective genus *Cichlasoma*. All *Thorichthys* cichlids need very similar care. One example is *T. aureus*.

Cichlids from South America

There are big differences among the cichlids from South America. Some, like the *Apistogramma* species, are tiny, whereas others, like the genus *Cichla*, are regular giants. There are open-water spawners among them, like *Pterophyllum scalare*, hidden brooding dwarf cichlids, and mouthbrooders related to the "earth eaters." Most imported cichlids from South America come from the Amazon region, and almost all of them prefer soft, slightly acid water, though they will adapt to neutral water values in a tank. This is also true of the "king" of cichlids, the discus fish or *Symphysodon aequifasciatus*.

Apistogramma cacatuoides

Cockatoo Dwarf Cichlid

Photo on page 56.

Geographic origin: Peru.

Size: Females up to 2 inches (5 cm); males 3 to 3½ inches (8–9 cm).

Appearance and sexual differences: Different color varieties. Males with elongated tips of fins, especially the first rays of the dorsal fin. Selectively bred fish often with bright spots on upper part of tail fin. Females are grayish yellow or, during brood care, yellow with dark longitudinal stripes.

Behavior: Peaceful; no threat to plants; also suitable for community tanks.

Care: Always place only one male with three to four females in a well-planted aquarium for South American cichlids (aquarium type 1 or 2, see page 12); well suited to join other, not too large tropical fish. Provide caves for spawning.

Water: 75° to 82°F (24–28°C); pH 6.5–7; 5–15°dH.

Feeding: Primarily live and frozen food; doesn't like dry food.

Breeding: Males form harems of several females with whom they spawn in turn. The water in the breeding tank should have the same properties as the regular tank water. These cichlids are hidden brooders and like clay flower pots for spawning. Males do not share in brood care but guard an extensive territory; females care very actively for their brood and guide the fry. This is the most undemanding species of the *Apistogramma* genus.

My tip: When buying this cichlid, make sure the males have good coloring. Maintaining the correct water quality is important.

Species with similar requirements: Some of the other *Apistogramma* cichlids are not too difficult to keep either, such as *A. borelli* (formerly *A. reitzigi*) and *A. agassizii*. But these as well as all other *Apistogramma* species are more sensitive to water quality, requiring a pH of 6–6.5 and 4–18° dH. For breeding, the hardness should in many cases be as low as 6°dH and the pH below 6. On the whole, the dwarf cichlids of this genus have to be considered sensitive (high standards for water quality, susceptibility to disease, sensitivity to medication).

Nannacara anomala is less fussy about water quality both in the regular and the breeding tank and is therefore a better bet for beginners interested in dwarf cichlids.

"Geophagus" steindachneri

Redhump Geophagus

Also known as *"Geophagus hondae"*

Geographic origin: The Guiana countries.

Size: Up to 8 inches (20 cm); tank-bred fish rarely over 5 inches (13 cm).

Appearance: Steep facial profile with, in males, a red hump on the forehead, which can change color. Irregular black spots on sides. Different local color varieties.

Sexual differences: Males larger, with hump on forehead.

Behavior: Peaceful, even toward smaller fish; however, this species likes to burrow.

Care: Community tank at least 40 inches (100 cm) long (aquarium type 2, see page 12); dense background planting recommended, but plants need to be well anchored; rocks and bog wood as decoration. One to two males with four to five females.

Water: 75° to 82°F (24–28°C); pH 6.5–7; 5–15°dH.

Feeding: All types of food described; prefers live food.

Breeding: Mouthbrooders, no pair bond; short courtship and spawning behavior typical of mouthbrooders (see page 40); spawning on stones. Mouthbrooding for 15–20 days, and the fry still are picked up in the mouth for some time thereafter. Breeding this species is not difficult.

My tip: Be careful if you change the water while the mother has eggs in her mouth; she might spit them out at the wrong moment.

Species with similar requirements: Members of the *Geophagus surinamensis* group, sometimes sold as "Surinam Pearls," grow to about 10 inches (25 cm). These are more placid fish, and there are larvophile as well as ovovphile mouthbrooders among them (see page 40). The same is true of the *Satanoperca* cichlids of the *"Geophagus" jurupari* group. *Biotoma* cichlids (photo on back cover) can be kept the same way, too.

Laetacara curviceps

Flag Cichlid

Formerly: *Aequidens curviceps*

Geographic origin: Amazon watershed.

Size: Females up to 3 inches (8 cm); males somewhat larger.

Appearance: Longish oval, somewhat high-backed; steep forehead. Dealers sell both tank-bred fish and ones caught in the wild, and several color variants exist.

Sexual differences: Juveniles are practically impossible to sex, and there are no reliable signs in adult fish either; general rules (see page 9) are all one has to go on.

Behavior: Peaceful; doesn't bother plants.

Care: One to two pairs in a small, well-planted tank (aquarium type 1 or 2, see page 12); also gets along in a community with other, not too small tropical fish; regular water changes very important.

Water: 73° to 82°F (23–28°C); pH 6–7; 2–12°dH.

Feeding: All types of food described; prefers live food.

Breeding: Open-water spawners; no pair bond; spawn on flat rocks or on pieces of wood; produce up to 200 eggs. The eggs often are eaten.

My tip: A species that is very easy to keep; suitable for beginners.

Species with similar requirements: *L. dorsiger*, a very similar species with the same requirements, has more red on the belly.

Papiliochromis ramirezi

Ram or Butterfly Dwarf Cichlid

Photo on page 27.

Former names: *Apistogramma ramirezi, Microgeophagus ramirezi*

Geographic origin: West of the Orinoco, in Venezuela and Colombia.

Size: About 2¾ inches (7 cm).

Appearance: Very colorful, with lots of shiny dots on body and fins, a vertical band across the eye, a black spot on the side, and black on the first rays of the dorsal fin. Tank-bred fish sold at pet stores often are much less attractive; there is also a pure golden yellow variety.

Sexual differences: Juveniles hard to sex; in adults, the female has a reddish belly, and the anterior rays of dorsal fin are elongated in the male.

Behavior: Pair-forming, peaceful cichlid; doesn't harm plants.

Care: Up to two pairs in a small, well-planted community tank (aquarium type 1, see page 12); somewhat sensitive to water quality; doesn't grow very old (about 2 years).

Water: 75° to 84°F (24–29°C); pH just barely below 7; up to 10°dH.

Feeding: All types of food described; prefers live food.

Breeding: The water definitely has to be softer (4–6°dH) and more acid (pH 6.5) than in the maintenance tank; peat filtering recommended. Open-water brooders that develop pair bond; spawn on small stones or in pits.

My tip: Try to get juvenile fish from a breeder who lets the young fish grow up with their parents; tank-bred specimens available in pet stores often are artificially reared. Don't combine this species with other cichlids that are too dominant.

Popular Cichlid Species

Pterophyllum scalare

Angelfish

Geographic origin: Amazon watershed.

Size: Length up to 6 inches (15 cm); can reach a height of over 8 inches (20 cm).

Appearance: Disk-shaped with large dorsal and anal fins that end in long, pointed tips. Sides have transverse black stripes that vary in number and prominence and that continue on the fins. Many different strains have been bred, such as the marble, veiltail, gold, and blushing angelfishes, as well as albino varieties.

Sexual differences: Can be sexed only by experts; at spawning time the genital papilla is pointed in males, flatter and round in females.

Behavior: Peaceful; doesn't burrow; tends to form shoals and becomes territorial at spawning time.

Care: Communal tank (aquarium type 2, see page 12) that is at least 40 inches (100 cm) long and 20 inches (50 cm) high. Raise six to ten young fish together. A cover of floating plants is beneficial, and there should be no strong water flow. Combine only with placid fish.

Water: For maintenance: 79° to 82°F (26–28°C); pH 6.5–7.5; 5–12°dH. For breeding: Around 86°F (30°C); pH 6.5; 5°dH.

Feeding: Live and frozen food, but also likes dry food.

Breeding: Open-water brooders that spawn against vertical substrates (large plant leaves, wood, flat stones); several hundred to over 1,000 eggs are laid, and the hatched larvae are first reattached to the substrate and later hidden in pits. Both parents look after the fry.

Species with similar requirements: *Pterophyllum altum* grows somewhat larger, is even taller (*altum* = high), and is more difficult to keep—comparable to discus fish (below). Among the cultivated strains there are hybrids of the two species.

Symphysodon aequifasciatus axelrodi

Brown Discus

Photo on page 55.

Geographic origin: Amazon watershed.

Size: Up to 8 inches (20 cm).

Appearance: Disk-shaped, with large dorsal and anal fins. Brownish ground color with nine transverse stripes between the eye and the tail; horizonal blue markings only on forehead and nape.

Sexual differences: Can be sexed only by experts; at spawning time the genital papilla is pointed in males, flat and round in females.

Behavior: Very peaceful; doesn't burrow; lives in groups, from which pairs split off and establish their own territories only to mate and raise young.

Care: Community tank (aquarium type 2, see page 12) at least 40 inches (100 cm) long and 20 inches (50 cm) tall. Raise at least six young fish together; if the fish are full grown, about six to ten should live in a tank at least 60 inches (150 cm) long; planting should not be too dense. Discus fish should be combined—if at all—only with very placid fish. They are susceptible to hole-in-the-head disease (see page 34).

Water: For maintenance: 82° to 86°F (28–30°C); pH 6.5–7.5; 5–12°dH. For breeding: 86°F (30°C); pH 6–6.5; 0–5°dH.

Feeding: Live and frozen food, also for rearing fry; variety and good quality of diet are important. Dry food is not always taken, but food pellets definitely should be tried. Pet stores sell excellent food designed especially for discus fish.

Breeding: Open-water brooders that spawn against vertical substrates (wood, flat rocks, flower pots or conical vases made of clay). Generally 100 to 200 eggs are produced. Selective breeding often can be done only in an unplanted breeding tank. Maintaining correct water properties especially is important. Select pairs that have separated from the shoal to engage in courtship. The hatched larvae are "glued" back to the substrate, and once the fry swim freely, they at first feed only on a skin secretion that forms on the parents' sides (see drawing, page 31). Later the fry are given Artemia nauplii.

Species with similar requirements: Other discus species and various selectively bred strains are available, but they are more difficult to keep (see the lower values of water hardness and pH) and breed. Another species that needs similar conditions is *Pterophyllum altum* (deep angelfish), the largest angelfish. However, the latter is much less hardy and easy to care for than the various strains of *Pterophyllum scalare* that routinely are available at pet stores.

Cichlids from Asia

Only two cichlid species come from Asia. Both belong to the genus *Etroplus*, and one of them is a widely kept aquarium fish.

Etroplus maculatus

Orange Chromide

Geographic origin: Sri Lanka and southern India (often in brackish water).

Size: Up to 3 inches (8 cm).

Appearance: Laterally very compressed and high-backed. Ground color yellow, with contrasting black markings that become much more pronounced during brood care. An all-yellow variety also is sold.

Sexual differences: No reliable clues for telling the sexes apart except for the blunt genital papilla the female develops before spawning. Females are sometimes a little less brightly colored, especially the reddish markings.

Behavior: Quite peaceful and nondestructive of plants; however, very few plants will grow in a tank with brackish water. In nature, this species serves as cleaner fish to the larger *E. suratensis*.

Care: Keep one pair (let pairs select themselves within a shoal of juveniles) in a tank (aquarium type 1 or 2, see page 12). In larger tanks two pairs can be kept together.

Water: 77° to 82°F (25–28°C); pH above 7; 5–30°dH. 1–2 tsp. of salt can be added per 10 quarts (10 l) of water, but this is not necessary.

Feeding: Live, frozen, and dry food.

Breeding: Parent-type family; eggs are deposited on rocks or roots, the larvae are then moved to pits. Intensive brood care, which is shared evenly by males and females.

Species with similar requirements: The banded chromide (*E. suratensis*) is similar but considerably larger (over 10 inches [25 cm] and therefore needs a larger aquarium (at least 48 inches [120 cm] long).

Addresses and Bibliography

Books

Hawkins, A.D., Editor. *Aquarium Systems*. Academic Press, 1981.

McInerny, Derek and Geoffrey Gerard. *All About Tropical Fish*. Facts On File, Fourth Edition, 1989.

Scheurmann, Ines. *The New Aquarium Handbook*. Barron's, 1986.

Sterba, G. *The Aquarists' Encyclopedia*. Blandford, 1983.

Ward, Brian. *The Aquarium Fish Survival Manual*. Barron's, 1985.

Magazines

Aquarium Fish Magazine
P.O. Box 6050
Mission Viejo, CA 92690

Freshwater and Marine Aquarium
144 West Sierra Madre Boulevard
Sierra Madre, CA 91024

Practical Fishkeeping Magazine
RR 1, Box 200 D
Jonesbury, MO 63351

Tropical Fish Hobbyist
One TFH Plaza
Third and Union Avenues
Neptune City, NJ 07753

Societies

American Cichlid Association
P.O. Box 32130
Raleigh, NC 27622

Index

Boldface type indicates color photographs. **C1** indicates front cover; **C2;** inside front cover, **C3;** inside back cover, **C4;** back cover.

Index

Index